REGULATORY INCENTIVES

FOR DEMAND-SIDE

MANAGEMENT

American Council for an Energy-Efficient Economy
Series on Energy Conservation and Energy Policy

Series Editor, Carl Blumstein

Energy Efficiency in Buildings: Progress and Promise
Financing Energy Conservation
Energy Efficiency: Perspectives on Individual Behavior
Electric Utility Planning and Regulation
Residential Indoor Air Quality and Energy Efficiency
Energy-Efficient Motor Systems:
A Handbook on Technology, Program,
and Policy Opportunities
State of the Art of Energy Efficiency:
Future Directions
Efficient Electricity Use: A Development Strategy for Brazil
Energy Efficiency and the Environment: Forging the Link

REGULATORY INCENTIVES

FOR DEMAND-SIDE

MANAGEMENT

Edited by
STEVEN M. NADEL
MICHAEL W. REID
DAVID R. WOLCOTT

American Council for an Energy-Efficient Economy
Washington, D.C. and Berkeley, California
in cooperation with
The New York State Energy Research and Development Authority
Albany, New York
1992

Regulatory Incentives for Demand-Side Management

Published by the American Council for an Energy-Efficient Economy,
1001 Connecticut Avenue, N.W., Suite 801, Washington, D.C. 20036
and 2140 Shattuck Avenue, Suite 202, Berkeley, California 94704.

Cover art copyright © 1938 M. C. Escher Heirs/Cordon Art — Baarn — Holland

Cover design by Chuck Myers, American Labor Education Center, Washington, D.C.
Book typeset by Wilsted & Taylor
Printed in the United States of America by Edwards Brothers, Inc.

Library of Congress Cataloging-in-Publication Data

Nadel, Steven, 1957
Regulatory Incentives for Demand-Side Management/edited by Steven M. Nadel, Michael W. Reid, David R. Wolcott.
    302 p.   23 cm.
   Includes bibliographical references and index.
   ISBN 0-918249-16-3 : $28.00
   1. Electric utilities—Government policy—United States. 2. Electric Utilities—United States—Energy Conservation.   I. Reid, Michael W., 1953–   II. Wolcott, David R., 1949–   III. American Council for an Energy-Efficient Economy
333.79'32'0973—dc20   92–28968
                                                CIP

NOTICE

Acknowledgments

We gratefully acknowledge our respective organizations for supporting our efforts to organize this volume and allowing us the time that was required to edit it. Barakat & Chamberlin generously contributed Mike Reid's services *pro bono*. The New York State Energy Research and Development Authority (NYSERDA) funded the research and production of Chapters 4, 8, 9, and 12 and partially funded the book's publication. The American Council for an Energy-Efficient Economy (ACEEE) funded the remaining costs, with significant support from the John D. and Catherine T. MacArthur Foundation and the Energy Foundation.

We could not have completed this work without the assistance and cooperation of a number of people. We first must thank each of the chapter authors for working with us and putting up with our editorial and schedule requirements. Tamaria Eichelburg at NYSERDA then did a skillful job of copy-editing. Victoria Pollicina of ACEEE made most of the copy corrections. Glee Murray, ACEEE's director of publications, managed the book's publication.

Finally, we wish to thank several individuals for inspiration and support for our work on DSM incentives. John Chamberlin and Julia Brown of Barakat & Chamberlin have provided valuable counsel in tracking and analyzing incentive developments. David Moskovitz and Jim Cole, now executive director of the California Institute for Energy Efficiency, catalyzed NYSERDA's interest in incentives which, among other accomplishments, led to support for this book. ACEEE executive director Howard Geller and ACEEE board president Carl Blumstein saw the need for this volume and supported its undertaking.

Stephen M. Nadel
Michael W. Reid
David R. Wolcott

Washington, D.C.
May 1992

Preface

Howard S. Geller, *Executive Director, American Council for an Energy-Efficient Economy*

Gunnar E. Walmet, *Program Director, New York State Energy Research and Development Authority*

Utilities in the United States, Canada, and other parts of the world are rapidly expanding their investment in end-use efficiency measures. In the United States alone, utilities are now spending around $2 billion per year representing approximately 1% of their revenues on demand-side management (DSM). The most aggressive utilities are investing 2% to 6% of their gross revenues in DSM. This growing commitment is being stimulated to a large degree by regulatory incentives.

As a strategy for increasing utility investment in end-use efficiency and thereby reducing the total cost of energy services, regulatory incentives move beyond trying to coerce utilities or to appeal to their sense of public duty. Rather they attempt to establish a system of rules whereby DSM investments are at least as profitable if not more profitable, when successfully executed, as conventional supply-side investments. The apparent logic of this approach has prompted more than 20 states to adopt or experiment with regulatory incentives since 1989. Although the incentive approach is still in its infancy, it is time to review its evolution, compare the different incentive options, examine preliminary results, and highlight problems and challenges. That is the objective of this pioneering book.

Top experts from across the country contributed chapters to this volume. The result, in our view, is a balanced presentation including utility, regulatory, and research perspectives. The book also features the latest initiatives and ideas related to regulatory incentives. It should be useful to those looking for an introduction to the subject as well as to those already working in the field. We hope that the book will lead to more extensive and effective regulatory incentive programs. At a minimum, it should broaden the understanding and appreciation of an important trend in the utility industry.

This book is the fourth major product of the collaboration between our two organizations that began in 1988. Earlier volumes analyzed electricity conservation potential in New York, lessons learned from utility DSM programs throughout the country, and programs for maximizing DSM implementation in New York. These state-of-the-art studies are helping to improve energy efficiency efforts in New York and throughout the country. We hope this volume will have similar effects, and we commend the editors and contributors for their excellent work.

Foreword

All ratemaking is incentive ratemaking. It rewards some patterns of conduct and deters others.

Many discussions of demand-side management "incentives" in ratemaking assume that no incentives existed until recently. This mistake has given rise to artificial debates about whether all utility demand-side management is being "subsidized."

Wise regulation understands the incentives that are woven into its ratemaking decisions and assures that they are consistent with sound management and state policy. Unsound regulation resorts to the historic formulas and incantations, treating the resulting utility behavior as something to be chastised or rewarded separately or—worse yet— legislated.

Anyone wanting to know the incentives implicit in historic ratemaking practice need only look at what the utilities did and didn't undertake—the good and the bad.

If managements have been mediocre in a particular jurisdiction for a long period of time, it is because ratemaking has neither rewarded good management nor penalized its opposite. If a particular jurisdiction has seen a frequent refusal of utility management to reevaluate commitments to unsound construction projects, that jurisdiction probably required that plant be "used and useful" before it could earn. If fuel purchase practices have been unsound, the villain is very likely to be the lack of scrutiny inherent in automatic fuel adjustment clauses.

Similarly, utility reluctance to become involved with demand-side management or with energy efficiency generally was itself the product of several incentives woven into conventional ratemaking.

First, until the nuclear construction debacles, utilities grew by increasing their investment in plant.[1]

This growth led to increased earnings and increased corporate and executive prestige. It also, in a time of substantial economies of scale

---

[1] This is the Averch-Johnson-Wellisz (AJW) effect discussed in Kahn, *The Economics of Regulation: Principles and Institutions*, Volume II, pp. 49–59, John Wiley & Sons, Inc., 1971.

in electric power generation, tended to lead to lower prices. These price reductions often took the form of incentives to consume, such as lower rates for high levels of use, or specific discounts for all-electric homes.

In addition, the tax incentives of the early 1970s—accelerated depreciation and the investment tax credit—biased utility decision making toward investment rather than conservation.

Fuel adjustment clauses compounded this tendency. Not only did they assure automatic recovery of the largest single variable cost, but their reconciliation mechanisms often assured that kilowatt-hours sold during peak periods would be profitable even though their price was below their cost of production.

Against this background, treatises in the 1970s by the Ford Foundation's Energy Policy Project[2] and by Amory Lovins[3] suggesting that the consumer's interest was in fact better served by conservation were widely unwelcome among utility executives, who spent more money in the 1970s seeking to discredit Lovins than they did on demand-side management.

As states began to require utility involvement in energy conservation, the interests of stockholders and customers diverged, a sure recipe for profound problems. The result was a sullen stalemate in which utilities praised conservation while damning conservationists. The required sums of money went into pilot projects, studies, conferences, and load shifting—but rarely into programs designed actually to save kilowatt-hours.

Early proposed reforms did not address the fundamental problem. The recovery of conservation expenses does nothing to diminish the incentive to sell kilowatt-hours. Inclusion of conservation costs in rate base creates incentives to institute ineffectual programs, which would then both earn a return and permit sales to continue.

Only in California was an adjustment mechanism (ERAM) adopted that actually assured that utility earnings would not suffer through declining sales. Not surprisingly, California utilities led the nation in their commitment to energy conservation in the early 1980s.

Fortunately, the last four years have seen the end of the stalemate in which public policy pointed in one direction and utility shareholder interests pointed in another. As a result, we now have sufficient experience with a range of demand-side management incentive programs to begin to develop a literature on the subject.

---

[2] *A Time to Choose: America's Energy Future*, Ballinger Publishing Co., Cambridge, Mass., 1974.

[3] "Energy Strategy: The Road Not Taken," *Foreign Affairs*, October, 1976.

Many difficult issues remain. The role of the federal government can charitably be described as undefined. Washington still prefers to focus its energy policy attention on issues that divide us—such as the Arctic Wildlife Reserve and nuclear power—instead of on issues that enjoy broad support, demand-side management and environmental considerations foremost among them. Striking a sensible balance between utility demand-side management programs and the demand-side initiatives of competing companies, especially in a time of recession, will inevitably lead to frustration before any real consensus is reached.

Issues of measurement and of comparison with supply-side alternatives abound.

The good news is that we have had scant cause so far to reconsider the fundamental commitment to reforming ratemaking in order to reward sensible demand-side management programs. If the fundamental ideas discussed in this book were wrong, the horror stories would already be beginning to accumulate. Certainly enough hostility and skepticism about demand-side management programs still exist to assure that the failures will be highly publicized. This compendium of articles helps to fulfill the commitment of every state commission and utility to learn from demand-side management operating experience. Only through such efforts can utility regulation be harmonized with customer response and utility management practices.

Peter Bradford
Chairman, New York
Public Service Commission

Contents

Chapter 1   Why Regulatory Reform for DSM *by David Moskovitz*   **1**

Chapter 2   The Evolution of DSM Incentives *by Michael W. Reid*   **21**

Chapter 3   California's ERAM Experience *by Chris Marnay and G. Alan Comnes*   **39**

Chapter 4   Revenue-per-Customer Decoupling *by David Moskovitz and Gary B. Swofford*   **63**

Chapter 5   Ratebasing of DSM Expenditures *by Michael W. Reid*   **79**

Chapter 6   Sharing the Savings to Promote Energy Efficiency *by Joseph Eto, Alan Destribats and Donald Schultz*   **97**

Chapter 7   Revenue Decoupling Plus Incentives Mechanism *by L. Mario DiValentino, Terry L. Dittrich, James E. Cuccaro, and Alan M. Freedman*   **125**

Chapter 8   Bill Indexing *by David Moskovitz and Richard Rosen*   **141**

Chapter 9   Utility Energy Services *by Charles J. Cicchetti and Ellen K. Moran*   **163**

Chapter 10   Evaluation of DSM Programs and Financial
              Incentives *by Eric Hirst*   187

Chapter 11   Spare the Stick and Spoil the Carrot: Why
              DSM Incentives for Utility Stockholders
              Aren't Necessary *by Paul Newman, Steven
              Kihm and David Schoengold*   203

Chapter 12   Does the Rat Smell the Cheese? A
              Preliminary Evaluation of Financial
              Incentives Provided to Utilities *by Steven M.
              Nadel and Jennifer A. Jordan*   229

Chapter 13   DSM Incentive Mechanisms: Comparative
              Assessment and Future Directions *by David
              R. Wolcott and Steven M. Nadel*   255

About the Editors   271

About the Authors   273

Index   277

Chapter 1

Why Regulatory Reform for DSM

David Moskovitz

Introduction

The United States has vast opportunities to increase energy efficiency. Greater reliance on energy efficiency can, in turn, reduce our national energy costs, increase our competitiveness and security, and improve our environment. Credible estimates indicate that cost-effective technologies available today have the potential to cut the nation's electricity use by 30% to 75% without lifestyle changes or reduced growth of the Gross National Product (GNP) (Gellings et al. 1990).

Much of this country's opportunity for reaching this higher level of energy efficiency lies with our regulated electric and gas utilities, the sector of the economy where energy decisions are most influenced by government control. Unfortunately, regulation as currently practiced in most areas of the country discourages rather than encourages the development of this important potential asset.

A carefully targeted reform of some aspects of current regulatory practice could accelerate development of this large energy efficiency resource. Without such regulatory reform, the energy efficiency resources available to the electric and gas industries will not be fully tapped, reducing the economic and environmental benefits that would otherwise accrue.

Of the ten industrialized nations that comprise the Organization for Economic Co-operation and Development (OECD), the United States ranks ninth in energy efficiency. We use twice as much energy as Japan, West Germany, or Sweden to produce a dollar of GNP (Flavin and Durning 1988). Only part of this difference in energy intensity can be explained by societal factors unrelated to national levels of energy efficiency.

Adopting cost-effective energy efficiency as the nation's energy investment strategy could substantially reduce the United States' annual $170 billion electricity bill. A cost savings of this magnitude would improve United States global competitiveness and would help reduce our trade deficit (Gellings et al. 1990).

Detailed macro-economic studies performed in conjunction with the National Energy Strategy show that relative to a business-as-usual scenario that does not emphasize energy efficiency, higher levels of commitment and investment in energy efficiency would result in a cleaner environment, produce increases in the nation's GNP, and reduce energy prices for American consumers (US DOE, 1991).

The environmental effects associated with electricity production are substantial. In the United States, electric utilities generate 20% of the gases linked to potential climatic change as well as 70% of sulphur dioxide and 33% of nitric oxide emissions. These latter two pollutants are primary contributors to acid rain and urban smog. In addition to atmospheric emissions, 85% of all nuclear waste is produced in electric generating facilities. Increasing the efficiency of our energy use, particularly electricity, can produce substantial environmental and health benefits (Ketcham-Colwill 1989, MacKenzie 1989).

Least-Cost Planning

In recent years many state regulatory agencies have made attempts to "capture" some of this potential resource. In the past few years, utility regulators in many jurisdictions have required utilities to use a least-cost planning (LCP) process when selecting new energy resources. In addition, a number of utilities have voluntarily embraced LCP.

LCP, sometimes called integrated-resource planning, requires utilities to consider energy efficiency resources equally with traditional supply-side resources in determining the best way to meet forecasted energy demand. The choice is based on cost; the less costly resources are selected for development before more expensive ones. In this manner, ratepayers and society will be able to meet future demands for energy while minimizing the total resource cost.

Because energy efficiency is often the least costly resource available, this selection process typically results in a sizable investment by our electric and gas utilities in energy efficiency. For example, in New England, Wisconsin, and the Pacific Northwest—where LCP practices have been used for some time—utilities are currently investing 3% to 6% of gross revenues in demand-side activities (Moskovitz, Nadel & Geller 1991). This level of investment is ten times greater than the level in most other states. Yet even in these high-investing regions, demand-side programming has not begun to exhaust available cost-effective energy efficiency opportunities.

Unfortunately, while a few utilities are genuinely pursuing least-cost planning, many are simply not investing in the efficiency resource. The average utility in the United States spends very little money on energy efficiency and focuses instead on load growth, load shifting, power marketing, or discount and incentive rates (Moskovitz, Nadel & Geller 1991).

Despite the strong interest of utility regulators in the LCP process, LCP has had a difficult transition from theory to practice. **The major obstacle to this broader application is related to traditional regulatory practices in most jurisdictions**.

Conventional regulation provides powerful disincentives to utility investment in energy efficiency and equally powerful positive incentives for utilities to *increase* sales. Conventional regulation links the earnings of utilities to their level of kilowatt-hour sales. Because energy efficiency *decreases* sales, this fundamental orientation of current regulatory practice penalizes energy efficiency investments and is contrary to the intent and goals of LCP.

Utility earnings from the demand side are also adversely affected by regulatory cost-recovery practices. Often these mechanisms produce an implicit bias toward the supply side because they sometimes fail to provide full and prompt recovery of utility demand-side resource costs.

Unfortunately, full development of the cost-effective energy efficiency resource in the United States will probably not occur without removal of these serious regulatory impediments.

Utility executives react to the regulatory and financial environment, and they respond with formidable financial, technical, legal, and political resources. The critical question for regulators, policy makers, and lawmakers is: can existing regulatory practices be successfully and equitably modified to create an environment that causes electric utilities to fully embrace and seriously implement least-cost planning?

Regulatory Impediments to Energy Efficiency Investment

The ratemaking process currently used in most jurisdictions provides these economic incentives for utilities:

- Regardless of its generation cost or price, each kilowatt-hour (kWh) a utility sells adds to earnings.

- Each kWh saved or replaced due to improved energy efficiency reduces utility profits irrespective of the cost of the demand-side measure.

- The only direct financial consequence of regulation that encourages utilities to pursue cost-effective conservation is the risk that dissatisfied regulators may disallow costs.

Clearly, these incentives are inconsistent with significant levels of utility investment in conservation. Admittedly, these existing incentives were not a conscious creation of the regulatory process as it evolved over the last century. Nevertheless, they are real and continue to exert a powerful and persistent influence over utility business objectives and corporate cultures.

Only limited progress implementing large-scale efficiency programs is possible in an environment dominated by such powerful, opposing economic forces. Although regulators rightly insist on the implementation of least-cost planning, the reality is that they administer a process that rewards utilities for non-conformance with that goal. Because utilities are "profit maximizing" entities, LCP is likely to find limited real success until means are found to align the financial interest of the industry with the goals of least-cost planning.

What is it about the traditional rate-setting process that produces all the wrong incentives?

Profits Are Not Fixed

As regulated monopolies, utilities are entitled to have their prices set at a level that allows recovery of all prudently incurred operating expenses and fixed costs, plus a reasonable rate of return on their rate base, calculated as their capital investment in power plants, transmission and distribution facilities, meters, trucks, inventory, and working capital, less depreciation.

However, contrary to public perceptions, utility profit levels are not fixed, capped, or guaranteed by regulators. Instead, at the time of formal rate cases, state public utility commissions set prices at levels to allow utilities to collect enough revenue to cover costs and earn a fair rate of return. The ratesetting process is *based on the assumption* that the relationship between future costs and sales levels will remain the same as that calculated by the commission. Unfortunately, this assumption about the future cost/sales relationship is rarely, if ever, borne out. From the moment prices are set at the conclusion of one rate case, to the moment they are reset at the conclusion of the next rate case, the utility has an incentive to sell more energy whenever its mar-

ginal revenue from the sale of power exceeds its marginal cost to produce and distribute that power.

The concept is simple. If you can sell a product for more than it
costs to produce, you make a profit. The more of the product you sell,
the more profit you make. This simple concept, however, becomes
complex when applied to electric utilities. Utilities provide services in
a monopolistic environment in which prices are set by regulators
instead of by the free market. Various regulatory instruments, particularly fuel adjustment clauses, purchased power clauses, and regulatory
accounting practices combine to assure that the cost *to the utility* of
producing more power is essentially zero. This net zero cost to the utility, as distinguished from the actual cost of the new power, is the result
of the "pass through" of the entire fuel cost and other variable costs
directly to the consumer. If the marginal cost of power to the utility, or
more precisely to utility shareholders, is essentially zero, every sale is
profitable.

If profits rise too high, regulators can intervene and lower the utility's price, but even when rates are lowered, the utility is not required
to give refunds or credits to customers to make up for previous excess
earnings. Thus, a utility may retain all the profit it earns between formal rate cases. Nor does having more frequent rate cases resolve the
dilemma: the utility is always "between" rate cases—the intermediate
period when there is an incentive to maximize sales.

Fuel Adjustment Clauses, Purchased Power, and Accounting Practices

The prevalence of fuel adjustment clauses is a major reason why the
cost of new power supply is not borne by utilities. Some 40% to 50%
of the average cost of electricity is determined by the cost of fuel (U.S.
DOE, 1991). To insulate utility shareholders from the impact of fluctuating fuel prices on earnings, nearly all states allow utilities to adjust
prices to customers using fuel adjustment clauses, which transfer the
cost of fuel from the utility directly to ratepayers so that changing fuel
costs, as well as many other variable costs related to increased operations, do not affect profits.

In most jurisdictions, this "fuel adjustment" protection operates
whether a utility's total fuel bill increases due to rising prices or
because more fuel is consumed to satisfy an increased demand for
electricity. A utility that spends more on fuel to meet increased demand
can recover every cent of these increased costs by raising the price of
electricity to spread the extra cost among its customers. Conversely if
less is spent on fuel than projected, due to lower sales resulting from

an effective conservation program, the utility must pass all of the fuel cost savings to consumers through lower rates without realizing any cost savings for itself.

Fuel adjustment clauses combined with ordinary price setting and accounting rules allow utilities to make money *even when they sell power for less than it costs to produce!* For example, to meet increased demand during peak periods, a utility may have to operate a relatively inefficient diesel generator that consumes ten cents' worth of fuel to produce one kWh of electricity. The regulated price of power might be seven cents per kWh, which represents five cents in fixed costs and two cents allotted for the utility's "average" fuel costs. It may appear that the utility would suffer a loss when it generates power at a cost of ten cents and sells it for seven. But the utility can recover the extra eight cents (the generator's ten-cent fuel cost minus the two-cent average fuel cost) later by relying on the reconciliation requirements of the fuel adjustment clause to raise rates. In effect, the utility charges customers fifteen cents for the kWh, seven cents now and eight cents later through the reconciliation provisions of the fuel clause. Meanwhile, the five-cent non-fuel component of its seven-cent rate remains with the utility, contributing to its bottom line.

Fuel is not the only example of a cost that is not incurred by a utility when an additional kWh is sold. For example, when a utility contracts for purchased power, whether from another utility or from non-utility suppliers, the costs are generally treated like fuel costs. The cost of new purchased power supply resources is passed on to customers through various adjustment clauses. In this case the incremental cost of power to the utility's shareholders is again zero and each kWh sold contributes to the utility's basic profits.

Generally accepted regulatory and utility accounting practices can produce a similar effect. Deferred accounting practices widely used by utilities, for example, permit them to incur costs that are in turn accumulated and later recovered from customers. These deferred costs are not treated as current operating costs that offset current operating revenues. The result is that increased sales allow utility profits to grow because utilities keep the revenue associated with increased sales while deferring related costs for recovery from customers at a later time.

The Path to Regulatory Reform

The National Association of Regulatory Utility Commissioners (NARUC) recognized the fundamental conflict between the incentives inherent in existing regulatory practice and the LCP process in a July 1989 Resolution (NARUC 1989). In that resolution, which noted the significant ratepayer and societal benefits that could be realized from successful LCP, NARUC urged its members to consider the disincentives connected with the development of demand-side resources and to adopt ratemaking mechanisms to correct the problem. Regulators were urged to reform the regulatory systems so the successful implementation of a utility's least-cost plan would be its most profitable course.

A word of caution: effective regulatory reform and the implementation of DSM incentives are *not* synonymous. Regulatory reform encompasses the entire ratemaking and accounting process. Effective regulatory reform may or may not result from the simple addition of a specific DSM incentive plan or reward the type described in later chapters. Even the combination of a DSM incentive plan and a lost revenue adjustment or a decoupling plan may be counteracted by other aspects of a particular state's ratesetting process. The goal of regulatory reform is, simply put, to ensure that activities consistent with a utility's least-cost plan are also consistent with the financial interests of shareholders.

This book reviews in detail the types of regulatory reforms and incentives that have been either applied or proposed to solve this regulatory dilemma. It is important, however, to clarify the precise character of the objectives the reforms should be designed to accomplish. Throughout this book, four broad questions will be asked about each reform approach:

1. Does the ratemaking mechanism:

   - Align the utility's financial incentives with least-cost planning?

   - Decouple profits from sales? Will profits change as sales go up or down?

   - Provide a positive incentive for DSM, or does it merely remove disincentives?

2. Is the ratemaking mechanism performance-based and measurable? Can key indices used in the incentive formula be objectively determined without major difficulties?

3. Is the ratemaking mechanism:

   - Understandable? Can it be readily grasped?

   - Predictable? Will the utility know in advance that a specific accomplishment will produce a specific effect?

   - Administratively simple? It should not be overly complex or difficult to administer.

4. Does the ratemaking mechanism encourage other beneficial outcomes, such as minimizing costs to society and/or non-participants? Does it discourage undesirable outcomes, such as "gaming" or "cream skimming"?

These issues are discussed in more detail in the following sections. In considering the merits of any proposal, the complete proposal and each of the issues previously listed should be compared to existing regulatory practices. Every proposal, no matter how well conceived, will have weaknesses and produce its own unique set of perverse incentives. Rather than ask if the proposed plan is ideal, regulators and other concerned parties should ask: (a) if the new incentive structure is better or worse than the incentives in the current system; (b) whether utilities are likely to act on any of the perverse incentives; and (c) whether unreasonable utility actions theoretically encouraged by a new regulatory framework are easily detected and, hence, deterred. While the ultimate goal of regulatory reform is to formulate a plan that is fully consistent with LCP, perfection is unlikely. Instead, proposals that provide realistic and significant improvements over the status quo should be pursued.

Aligning Utilities' Financial Incentives with LCP

The most important objective of reform must be to align the utilities' financial incentives with LCP. To identify the approaches that meet this basic objective, reform plans can be tested using a simple series of questions. For example:

- Viewed from the utility's perspective, what strategy would maximize profits?

- What happens to profits if the utility sells another kWh?

- What happens to earnings if sales are reduced by one kWh through conservation programs that cost one cent per kWh?; two cents?; ten cents?

- What happens to profits if a utility invests in load control and shifts a kilowatt from on-peak to off-peak? What happens if the utility pursues a power-marketing strategy?

- What happens if the utility selects the more costly of two supply-side options; or the more costly of two demand-side options; or a supply-side option that is more costly than a demand-side option?

Answering these questions requires knowledge of the specific ratemaking and accounting practices used in the state, particularly:

1. The precise workings of fuel and purchased power clauses and associated reconciliation provisions;

2. Ratemaking provisions allowing deferred expense accounting, including deferred accounting for conservation costs; and

3. Rate levels and rate structures for each customer class and for each time period where time-of-use rates have been implemented.

The utility's most profitable course should be to successfully implement a least-cost plan. If the utility's most profitable course is to pursue programs that do not reflect a cost-minimizing plan while still promoting sales that are not cost-effective, the regulatory reform plan fails to meet this primary criterion.

Removing the Disincentives

Decoupling Profits from Sales. Under current regulation, increased sales always mean increased profits, while lower sales due to energy efficiency improvements or other causes invariably translates into diminished earnings. As long as these conditions exist, a strong likelihood remains that a profit-maximizing strategy will lead to more sales and less DSM, even if DSM programs are profitable.

Decoupling is the term used to describe a reform plan that breaks the linkage (or coupling) between profits and sales. Decoupling is the single most important step to removing existing disincentives, thereby encouraging the successful implementation of the least-cost process.

The most widely known decoupling mechanism is California's Electric Revenue Adjustment Mechanism (ERAM), discussed in Chapter 4. ERAM-type mechanisms, however, rely on future-test-year ratesetting, a practice used in only a few states. Another decoupling plan, based on established fixed revenue per customer has recently been developed and implemented in Washington, for Puget Power and Light Company, and in Maine, for Central Maine Power Company. This approach, which can be used in both future- and historic-test-year jurisdictions, is discussed in Chapter 5.

There are also other, very different, approaches that may accomplish similar results. For example, innovative fuel revenue accounting methods can be used to reduce the amount of non-fuel revenues a utility earns from marginal sales. If marginal non-fuel revenues were entirely eliminated, incremental sales would not contribute to additional profits. While accounting changes of this general type were adopted in Maine, they lacked the capability to decouple profits from sales (Moskovitz 1989).

Eliminating or substantially modifying fuel adjustment clauses, a step that would be resisted strongly by utilities, is at best only a partial solution to the coupling of profits and sales. With a typical fuel adjustment clause, electricity sales add to profits whenever prices exceed average fuel costs. Eliminating the fuel adjustment clause means profits rise when prices exceed short-run marginal fuel costs. The difference between the two situations is that with fuel clause reform, utilities no longer have an incentive to increase sales when marginal costs exceed prices. Currently, average and marginal fuel prices are about equal, and both are substantially less than average electricity prices. Thus, while fuel clause reform can provide needed incentives to control fuel and purchased power expenses, reforming or eliminating fuel adjustment clauses alone will not remove the existing disincentives to achieving energy efficiency.

It may also be possible to accomplish decoupling with a reward/ penalty mechanism. For example, financial rewards and penalties for utility programs that produce changes in average customer bills can decouple profits from sales. This decoupling effect, however, can only occur if the rewards and penalties are large enough to counter the increased revenue earned from the sale of power. This approach is discussed in Chapter 8, Bill Indexing.

Lost Revenue Adjustments. Many states attempt to remove the existing disincentives to demand-side investment using lost revenue adjustments. Under this approach utility revenue losses associated with approved DSM measures are estimated or measured and the utility is allowed to recover the revenues from customers. Unfortunately, reform plans relying on this type of adjustment mechanism do not remove the incentive to sell power and therefore do not break the linkage between profits and sales.

At best, lost revenue adjustments, when combined with rigorous after-the-fact energy savings measurements, can remove some DSM disincentives. However, this is true only with specific, readily tracked demand-side programs. Without decoupling, improvements in energy efficiency caused by utility educational programs, improved prices that induce more customers to invest in energy efficiency, or the legislative implementation of energy efficiency standards will continue to adversely affect the utility's shareholders' return. The predictable result is continuing utility reluctance to pursue a diverse range of energy efficiency options.

If lost revenue adjustments are combined with engineering or other preestablished savings estimates, perverse incentives are created. Estimated energy savings that are not realized reward the utility twice: once with the assumed lost revenues, and again with the revenue from kilowatt-hours that were not successfully saved (see measurement discussion later in this chapter).

In addition, lost revenue adjustments raise new measurement and policy issues:

1. How to treat increased sales to industrial or commercial customers whose improved energy efficiency has contributed to higher production levels;

2. How to treat lost revenues resulting from rate design improvements; and

3. Whether and how to offset lost revenues from efficiency gains with increased revenues resulting from strategic load building, valley filling, load management, or increased off-system sales.

If the existing sales incentives remain intact, a lost revenue adjustment may be insufficient to allow the necessary shift in utility focus and infrastructure to transform LCP into a major utility objective. Moreover, as more rigorous measurement and other adjustments are incorporated into lost revenue adjustments, utilities are likely to perceive an increased cost-recovery risk, with the result that utility managers will avoid DSM resources.

DSM Cost Recovery. Decoupling, and to a lesser extent lost revenue adjustments, remove the greatest barrier to utility investment in DSM. The remaining disincentive relates to the recovery of DSM costs. Under current regulatory practices utilities are assured full and prompt recovery of the cost of meeting customer energy service demand with supply-side resources. Fuel adjustment clauses and similar purchased power clauses assure full cost recovery for all prudent fuel and purchased power costs. Capital cost accounting, including accrual of interest and other costs of capital (Allowance for Funds Used During Construction, or AFUDC) for new power plant construction, assures that prudent supply-side resource costs are also fully recovered.

A disincentive to investment in DSM will exist if regulation allows recovery of DSM costs on terms less favorable than recovery of supply-side costs. DSM cost-recovery options are discussed in Chapters 2, 5, 7, 9, and 11.

Considerations for Positive Incentives

Removing disincentives to DSM is necessary, but it may not be enough to focus the attention of a utility's top management on LCP in general, or DSM in particular. Adding reasonable performance-based positive rewards or bonuses has produced impressive results (Rowe 1990).

Measurability

DSM program evaluation is discussed in Chapter 10. In addition to the need to assure satisfactory performance of DSM programs over their anticipated useful lives, evaluating the results is important because the success of any incentive will be greatly influenced by the character and timing of the energy savings measurements. Measurement considerations are not mere technical issues.

Energy savings estimates based on engineering or economic calculations instead of actual measurements are often adequate for some purposes. However, unless an incentive plan separately decouples profits from sales, regulatory incentive proposals relying on engineering or similar estimates may produce a poor set of incentives.

To illustrate the interaction between measurement approaches and decoupling efforts, consider the substantially different incentives produced by an electric water heater insulation program under two plans where the only difference is how and when program savings are measured. The first plan has kWh savings based on extrapolating test data, engineering estimates, or measurements made at other times or in other states. The second plan is the same except that program savings are based on random, statistically valid, on-site measurements of utility-installed measures. Both plans allow the utility to recover direct rogram costs plus lost revenues associated with the program.

Suppose, under the first plan, an agreement is reached that an electric water heater insulation blanket will yield 600 kilowatt-hours per year in energy savings. Under this plan, the utility will be allowed to recover direct and indirect program costs, 600 kWh's worth of lost revenues, and an incentive based on any rational approach. In this example, the exact nature of the incentive element is unimportant.

What happens when the utility actually achieves 700 kilowatt-hours in savings through better quality control or other efforts under its control? Greater-than-estimated kWh savings cause it to lose money. An equally perverse result occurs when the utility selects inefficient contractors and actual savings drop to 500 or 400 kWh per year. Despite poor performance, utility profits increase because the utility still recovers lost revenue based on an assumed 600 kWh savings, although not all these revenues were lost. In addition, the incentive portion is unaffected by the lower actual savings. Thus, solely as a consequence of a measurement decision, the utility's profit-maximizing strategy would be to select measures that would test well using the measurement criteria imposed but perform poorly in practice.

Note that this would not be the result if the plan decoupled profits from sales instead of allowing recovery of lost revenues. With decoupling, consumers, rather than the utility, would receive the excess revenue from the higher sales resulting from poor DSM performance.

With decoupling it may be reasonable to distinguish DSM evaluation for the purpose of program design, from DSM evaluation for the purpose of cost recovery and incentives. Using detailed and rigorous post-program evaluations as the primary basis for cost recovery and incentives places more financial risk on a utility than if a traditional supply-side resource been selected.

Understandability

Many different interests are represented in most regulatory proceedings. Not only are utility shareholders and ratepayers present in most proceedings, but more special interests—business customers, low-income customers, the environmental community, labor, and other groups—are often represented as well. Significantly reforming a regulatory system that has been in place for nearly a century requires substantial public and political support from as many of the interests affected as possible. Gaining the needed support will be difficult if the proposed plan is too complex or obscure. For this reason, reforms should be easily understood by all parties.

Predictability

While regulators will always have great discretion in ratesetting proceedings, incentive proposals that give explicit guidelines—so utilities know that a specific action will result in a particular gain or loss—will motivate utility managers better than alternatives that rely heavily on commission discretion.

Regardless of how responsible, consistent, and objective regulators may be, they will frequently be viewed as capricious by the regulated utilities or other parties. Even when there is no outright distrust, the relatively short tenure of most commissioners—about four years in the United States—adds to the lack of predictability of outcome for approaches that rely heavily on commission discretion. Consequently, incentive proposals related to the discretion of commissioners may not achieve full potential in motivating utility managers, even if the commissioners are reasonable and responsible.

Predictability does not mean that the utility should be guaranteed a particular level of earnings in advance. Rather, the utility must know a specific action or accomplishment will produce a particular and predetermined result. The greater and more immediate the cause and effect, the more likely the regulatory incentives will have a definite influence on utility managers.

Administrative Simplicity

Incentive plans should be simple and efficient to administer, or the cost of regulation may outweigh the benefit. Such regulatory costs include the commission's expense of administering the system, utility expenses for information collecting and reporting, and the cost to all parties of participating in any new regulatory proceedings that may be needed.

In practice, this principle means avoiding incentive plans that rely on complex formulas or unverifiable measurements. For this reason, policy makers may want to avoid approaches that require separate proceedings in favor of plans that can be implemented within the framework of existing regulations.

Cost Minimization

Will the proposed program encourage the utility to deliver conservation programs at the lowest cost to consumers?

Consider two incentive plans, both of which measure actual achieved conservation benefits. The first reimburses the utility with a predetermined, fixed amount for each kWh saved (i.e., a bounty on energy savings) and the payment is in lieu of any other DSM cost-recovery mechanism. If the payment is less than the utility's avoided cost, the utility's interests will help assure that only cost-effective DSM is pursued. Any difference between the bounty and the utility's cost of DSM is retained by the utility. The second plan pays the utility 110% of its actual program expenses for each kWh actually saved.

To maximize profits under the first plan, the utility will try to reduce its expense of saving kWhs to maximize the difference between the fixed payment it receives and its out-of-pocket costs. To maximize profits under the second plan, the utility would pursue as much conservation as it could, regardless of the price.

Generally, plans should be designed to encourage utilities to obtain DSM savings at the lowest reasonable cost. However, if program expenses are too low, program participation rates may suffer. Thus the cost minimization objective must be balanced against other objectives.

Non-Participant Impacts

Do the proposed regulatory reforms encourage the use of DSM programs that minimize non-participant impacts? Depending on the utility's average and marginal costs and the state's specific mechanisms for DSM cost recovery, DSM programs may have an adverse albeit small impact on average prices, thereby raising prices *and* bills for customers who do not participate in DSM programs. Rates for participating customers increase as well, but the DSM program reduces their bills (Hirst, 1991).

The non-participant impact of even an extremely large DSM program is minimal, much smaller than the impact of a typical supply-side option or the impact of routine regulatory decisions involving such matters as rate design and cost allocation (Cavanagh 1988). Nevertheless, incentive plans can be structured to encourage utilities to minimize non-participant impacts without jeopardizing customer participation rates. For example, plans that encourage utilities to minimize DSM costs will at the same time tend to minimize non-participation rates. Plans can also be designed to provide incentives for utilities to obtain as much contribution as possible from participating customers. As the proportion of DSM costs paid by customers increases, fewer dollars need to be contributed by ratepayers with the result that impacts on non-participants diminish. The primary risk of this approach, however, is that customer participation rates may suffer.

A few utilities are currently experimenting with a DSM delivery approach in which participating customers pay for the full cost of utility-sponsored DSM programs through an energy service charge. This approach is designed to lessen non-participant impacts. These ongoing efforts will determine whether this approach can successfully achieve significant penetration of DSM measures. It may be possible to establish the energy service charge at a level that also includes a utility incentive. Chapter 9 gives a more complete description of the energy service charge approach.

Finally, non-participant impacts may be addressed by assuring that energy efficiency programs are widely available to all customers and all customer classes to minimize the number of customers who don't participate.

Skimming the Cream

Will the proposed incentive plan encourage the utility to engage only in the least expensive efficiency programs and leave other more expensive, but still cost-effective, measures undone? If so, is this a concern?

As an example of cream skimming, heating and cooling retrofits at a given facility might cost four cents per kWh saved, when installed at the same time as two-cent lighting improvements. If installed separately, however, the same HVAC measures might cost six cents. An incentive program that pays the utility five cents for each saved kWh might cause the utility to improve the lighting and earn three cents while foregoing the four-cent cooling improvement that would have netted only one cent. An incentive plan paying the utility three cents for lighting and five cents for heating and cooling would net the utility the same penny for both projects. To counter concern that the utility might then pursue the easiest lighting and heating opportunities, bounty payments for low-cost lighting improvements could be set lower than payments for more expensive measures.

The best argument against cream skimming is that cost-effective opportunities will be permanently lost and consumers will overpay for future energy services. While the DSM opportunities at risk are cost-effective, the payback on the less cost-effective measures is below the hurdle rate for the investing entity.

A plan that suffered only from the potential for cream skimming would be a major improvement compared to the current system. Nevertheless, one should be aware of the possible problem and the available solutions apart from the incentive plan.

Correcting the existing perverse incentives is not a complete substitute for regulatory oversight or public participation in a least-cost planning process. Collaborative efforts in New England suggest that utilities, energy efficiency advocates, and others can work together to design conservation programs that avoid cream skimming.

Avoiding Gaming

Any regulatory system, including traditional utility regulation, is subject to efforts by individuals to engage in short-term "gaming." Simple manipulations, like timing rate case filings, or timing certain maintenance expenses that can be deferred or accelerated, can affect the utility's bottom line. Regulatory proposals should be carefully designed and selected so that the opportunity for gaming is no greater than that generated by current regulatory practices.

One way to reduce the incentive for manipulation is to assure that the implemented plan will remain in effect long enough to make gaming risky. In addition, short-term gaming temptations may be minimized by ratebasing DSM expenditures or otherwise structuring the capitalization and amortization of DSM program costs to relate to program benefits. This issue is discussed in Chapters 3 and 5.

Balance

Incentive proposals should have a reasonable risk/reward relationship. When measurement criteria are set, superior performance should yield higher earnings and inferior performance should yield lower earnings. The plan should not permit utilities to profit at the expense of ratepayers, or deprive them of an opportunity to earn a fair return.

To gain public acceptance and increase the possibility of producing the desired result, an incentive plan should operate symmetrically by rewarding superior, and punishing inferior, performance. Examples of several plans are discussed in Chapters 6 and 7. Incentive plans that reward utilities for good performance, and do nothing when performance is poor, are unfair and ineffective.

Scope

Ideally, an incentive plan will include both demand and supply aspects of LCP. Most proposals are currently limited to making DSM programs profitable and do not address the incentives in traditional regulation to increase sales or any aspect of supply-side options. This should come as no surprise because the existing incentives for DSM are most skewed.

Limiting the scope of the undertaking, however, may narrow the range of available options, and needlessly eliminate effective approaches for that fit well with ratemaking or accounting practices unique to the state. For example, an option that modifies elements of the fuel adjustment clause would affect both DSM programs and sales incentives. Narrowing the scope of incentive plans to only DSM incentives may needlessly eliminate using this type of approach.

Other Considerations

Other considerations may influence the design of an incentive plan. For example, will the plan reward, punish, or be indifferent to programs that achieve cost-effective fuel switching? Will the plan operate as intended if a utility incorporates environmental externalities in its planning process? These and other questions may require attention before a regulatory reform plan is adopted.

Conclusion

Crafting a "reformed" regulatory system as outlined in this introduction is obviously not simple. Evaluating multiple options and considerations requires a great deal of time and care, but it can and should be done. The remainder of this book provides a thorough background on the evolution of regulatory reform and describes some of the most successful and promising options. It is our hope that this material will be useful in clarifying issues and will serve as a guidebook for further regulatory reform.

Chapter 2 provides a history of incentive regulation to promote DSM. Chapters 3 and 4 describe mechanisms that remove disincentives through decoupling approaches. Chapter 5 focuses on the remaining disincentive, DSM cost recovery. Chapters 6, 7, 8, and 9 describe the principal methods of providing positive rewards for utility investment in DSM. Issues relating to DSM evaluation are explored in Chapter 10. The final chapters take another look at traditional regulation (Chapter 11), assess the effectiveness of incentives adopted before 1991 (Chapter 12), and review the strengths and weaknesses of different incentive approaches as well as take a look at the future of incentive regulation (Chapter 13).

References

Cavanagh, Ralph. 1988. "Responsible Power Marketing in an Increasingly Competitive Era." *Yale Journal on Regulation* 5(331). Summer. Pp. 331–366.

Fickett, Arnold, Clark Gellings and Amory Lovins. 1990. "Efficient Use of Electricity." *Scientific American* 263(3). September. Pp. 65–74.

Flavin, Chris and Alan Durning. 1988. "Building on Success! The Age of Energy Efficiency." *Worldwatch Paper* 82. Washington, D.C.: Worldwatch Institute.

Hirst, Eric. 1991. *The Effects of Utility DSM Programs on Electricity Costs and Prices.* ORNL/CON-340. Oak Ridge, Tenn.: Oak Ridge National Laboratory.

Ketcham-Colwill, J. 1989. "Acid rain: Science and Control Issues." *Environmental & Energy Study Institute Special Report*, July 12. Washington, D.C.

MacKenzie, James. 1989. *Breathing Easier: Taking Action on Climate Change, Air Pollution, and Energy Insecurity.* Washington, D.C.: World Resources Institute.

Moskovitz, David. 1989. *Profits and Progress Through Least-Cost Planning.* Washington, D.C.: National Association of Regulatory Utility Commissioners.

Moskovitz, David, Steven Nadel, and Howard Geller. 1991. *Increasing the Efficiency of Electricity Production: Benefits and Strate-*

gies. Washington, D.C.: American Council for an Energy-Efficient Economy.

National Association of Regulatory Utility Commissioners (NARUC). 1989. *Resolution in Support of Incentives for Electric Utility Least-Cost Planning*. Washington, D.C.

Rowe, John, 1990. "Making Conservation Pay: The NEES Experience." *Electricity Journal* 3(4). December. Pp. 18–25.

U.S. Department of Energy (DOE). 1991. *National Energy Strategy, First Edition, 1991/1992*. Washington, D.C.

Chapter 2

The Evolution of DSM Incentives

Michael W. Reid

Introduction

The subject of DSM incentives has moved from novelty status to the mainstream in the span of just a few years. As of late 1991, DSM incentives are in place or on the regulatory agendas of some 30 states. There is so much activity in this area that monitoring the number of states and utilities considering incentives is challenging; significant developments occur nearly every month.

These activities don't occur in isolation: as in other areas of utility regulation, state commissions and utilities depend on precedents and the experiences of other states when considering DSM incentives. The results are evolutionary; the issues addressed in incentive proceedings and the designs of the mechanisms often reflect the influence of developments in states that have been in the vanguard on incentives.

This chapter traces the evolution in theory and practice of DSM incentives. Major developments and decisions over a period of more than a decade are described, and important issues raised in incentives proceedings are highlighted. This chapter is, in part, a preview of some of the detailed descriptions and analyses provided in later chapters.

Early Precedents for DSM Incentives

Washington State

Although the majority of state actions on DSM incentives have occurred since 1989, precedents date back to at least 1980. In that year, Washington State enacted legislation directing that utilities be granted a 2% bonus rate of return on the equity portion of investments that were "reasonably expected to save, produce, or generate energy at a total incremental system cost . . . [that was] less than or equal to the [cost of energy from] conventional energy resources which utilize nuclear energy or fossil fuels" (Washington 1980).

As interpreted by the Washington commission, this statute authorized two types of incentive treatments. First, utilities would be allowed to treat their conservation program expenditures as investments, rather than period expenses. The costs of programs would be "ratebased," that is, capitalized and amortized (recovered as a cost) over a multiyear period, earning a return annually on the unamortized portion, similar to supply-side investments. Washington allowed one utility, Puget Sound Power & Light, to ratebase virtually all its conservation program expenditures during the 1980s, although most of the dollars were not used to acquire physical assets, such as the generators and transmission equipment that traditionally comprise the majority of the rate base.

Second, the Washington statute provided a bonus return or "kicker" on the conservation investment—a return two percentage points higher than that allowed on the other components of the rate base. This was apparently the first time preferential financial treatment was applied to DSM expenditures.

According to analyses of DSM ratebasing, the incentive effect of ratebasing per se is modest; theoretically, a ratebased DSM investment earning the same rate of return as other utility assets offers no greater financial benefit to utility shareholders. But ratebasing may appear advantageous to utilities that are concerned about the implied reduction in rate base when DSM expenditures that are ordinarily expensed substitute for supply-side investments that would have been capitalized. This point of view is taken up in Chapter 5.

While the 2% bonus offered by Washington would appear to represent a true incentive, experience there suggests that it did not significantly stimulate DSM. Incremental conservation efforts still penalized utilities, since the bonus was insufficient to offset the conservation-induced loss of revenue between rate cases. Further, there was no provision for utilities to accrue carrying charges on DSM investments made between rate cases. The investments would be added to rate base, and begin earning a return, only at the conclusion of a rate case. The net result was that supply-side investments remained financially more favorable to utilities than DSM (Blackmon 1991).

Despite shortcomings in the incentive concepts applied in Washington, ratebasing of DSM and bonus rates of return on DSM investments are recurring themes in recent incentive proposals.

California's ERAM

A second major incentive-related development of the early 1980s was the creation of ERAM, the Electric Revenue Adjustment Mechanism. ERAM was proposed in a 1981 California case involving Pacific Gas & Electric (PG&E). PG&E was in serious financial straits, caused primarily by continuing cost and technical problems with its Diablo Canyon nuclear plant. To improve its financial stability, PG&E proposed reconciling actual base (non-fuel) revenues to the revenue level authorized by the commission. The proposed ERAM mechanism, endorsed by the commission staff and other parties to the case, took effect in 1982. Over the next three years it was implemented for California's other major utilities.

By "trueing up" utility revenues to an authorized level, ERAM eliminates fluctuations in revenue for whatever reason, including energy conservation. A utility subject to ERAM does not lose authorized base revenue when it increases its DSM efforts between rate cases, because the shortfall is collected from ratepayers in the next period, with an appropriate adjustment for interest. Symmetrically, any gains from expanded sales are returned to ratepayers. This aspect of ERAM has come to be known as "decoupling" because it severs the link between base revenue and the level of sales. ERAM does not make DSM more profitable for utilities than other resource options, but it removes the short-term revenue penalty from DSM, and it makes sales promotion less attractive than under traditional regulation.

At the time of its enactment ERAM's effect on DSM was a secondary consideration. In time, stimulation of DSM and the desire to decouple utility revenues from sales eventually became the dominant reasons for California's retention of ERAM. Chapter 3 discusses ERAM in detail.

Mid- and Late 1980s Wisconsin Experiments

Another approach to incentives was established in the mid-1980s in Wisconsin, beginning with a Wisconsin Electric Power Company (WEPCO) rate case decided at the end of 1986. When WEPCO proposed refurbishing an outdated generating plant to meet growing demand, the commission directed the company to begin a large-scale "conservation construction program" instead, emphasizing rebates and low-interest loans to customers who invested in energy efficiency measures. To reinforce the analogy to the supply side and spread the costs of the program over several years, the commission authorized ratebasing most of the utility's outlays (Wisconsin PSC 1986).

Wisconsin rejected WEPCO's request to fix a premium rate of

return on its ratebased DSM, like that authorized in Washington. In the commission's plan, WEPCO's return on DSM would be increased *if* the company reached a target level for reduction in peak load. One percentage point additional return on the equity portion of DSM would be received for each 125 MW of demand reductions.

The WEPCO decision was the first to tie a financial bonus to quantified DSM performance. The Wisconsin commission subsequently adopted different financial incentives for other utilities under its jurisdiction, in each case designing the mechanism so that performance or cost-effectiveness, rather than the level of spending per se, would be rewarded. Some of these schemes are described in Chapter 11, which also explains why regulators' enthusiasm for incentives that benefit shareholders has diminished in Wisconsin, just as other states are beginning to experiment with such mechanisms.

1988: A New Focus on Curing the Disincentives

In the mid-1980s the term "least-cost planning" (LCP) was solidly established in the utility lexicon. LCP, also called integrated resource planning (IRP), was a new planning paradigm that promised to minimize the costs of energy services (light, thermal comfort, torque, etc.) by expanding the menu of possible resources to include demand-side and nontraditional supply-side options. By 1988, 17 states were practicing some form of IRP, and a majority of states were moving in that direction (Barakat, Howard & Chamberlin 1988).

LCP requires new methodologies for comparing the costs of resources. In this context, regulators and industry analysts began to focus on whether DSM and supply-side resources were competing on a level playing field—that is, whether existing methodologies were biased. Some observers went further and questioned whether it was the nature of regulation, rather than the analytical methodologies, that tilted the playing field in favor of traditional supply-side resources. If this were so, only wholesale regulatory reform, rather than refinement of planning methodologies, could produce the least-cost outcome.

Attention to this viewpoint increased considerably in April 1988 when David Moskovitz, at that time a member of the Maine commission, addressed the first national conference on least-cost planning sponsored by the National Association of Regulatory Utility Commissioners (NARUC). Moskovitz's speech included several provocative statements:

Without significant reforms to the ratemaking system, least-cost planning is going nowhere.

Least-cost planning will not work in most states because it is inconsistent with the type of economic regulation presently used in the utility industry.

There is no incentive for the utilities to encourage demand-side measures and every incentive not to encourage them, except perhaps as an appeasement to their commissioners. (Moskovitz 1988)

Shortly thereafter, NARUC's Energy Conservation Committee sponsored a workshop on the relationship between least-cost planning and utility profitability. The result was a committee resolution urging state commissions to

. . . adopt appropriate mechanisms to compensate a utility for earnings lost through the successful implementation of demand-side programs which are part of a least-cost plan and seek to make the least-cost plan a utility's most profitable resource plan. (Energy Conservation Committee 1988)

The NARUC resolution catalyzed several state regulatory commissions to begin studying the incentives issue. Some utilities also began to formulate proposals for financial incentives. By early 1989, the DSM incentives movement was in full swing, with an emphasis on identifying and curing the *dis*incentives affecting utilities' pursuit of DSM. The resulting proposals often had three components:

• A cost recovery mechanism to ensure that the utility would be able to recover in rates all prudently incurred costs of DSM programs.

• A lost revenue mechanism that would adjust rates to compensate for the short-term loss in base revenue that results when DSM programs succeed.

• A bonus provision, usually linked to performance, that would (1) help offset the risks that utilities often perceive in DSM, and (2) reward utility shareholders for cost-effective expansion of DSM programs.

Most of the incentive proposals currently being considered follow this three-part formula.

Trend-Setting Incentive Actions

Regulatory proceedings during 1989 and 1990 resulted in the establishment of several DSM incentive plans that have become the reference points for most subsequent action in this area. Among the most important developments were the New York State order establishing incentives for Orange and Rockland Utilities (O&R) and Niagara Mohawk Power Company (NIMO); the approval of parallel DSM incentive proposals put forth by New England Electric System (NEES) in three states; the development of incentive proposals for California's major utilities in a collaborative process; and the revision of Orange and Rockland's original incentive after less than a year of operation. Each of these developments is briefly described in this section.

New York's Orange and Rockland/Niagara Mohawk Decision

In 1988 the New York commission directed utilities to develop and file comprehensive DSM plans. At the same time, the utilities were invited to submit proposals to reform ratemaking "such that DSM programs that benefit customers are also rewarding to stockholders" (New York PSC 1988).

The commission dealt with the two best-developed proposals, those of O&R and NIMO, in a September 1989 order that established a pattern for several subsequent decisions. By July 1990, all seven investor-owned utilities in New York had DSM incentives in place. While the mechanisms were generally similar, none were identical; to foster innovation and gain experience with incentives, the commission permitted the utilities broad flexibility in the design of the mechanisms' details.

The NIMO/O&R decision authorized the utilities to estimate and collect from ratepayers DSM-related lost revenues, i.e., the portion of authorized base revenue that is foregone when DSM succeeds. Like ERAM, this adjustment was seen as a means to eliminate the financial penalty of expanding DSM programs between general rate cases.[1]

Further, the utilities were granted DSM bonuses, expressed as shares of the net savings resulting from selection of DSM in lieu of supply-side options. O&R, for example, was authorized to receive as shareholder profits 20% of the net benefits (gross benefits minus costs) resulting from DSM. The shared-savings approach linked the bonus to the utility's performance and was expected to motivate both expansion in the size of programs and efforts to maximize cost-effectiveness.

---

[1] Earlier in 1989 the New York commission had hastily adopted a lost revenue adjustment in a settlement involving Long Island Lighting Company. The O&R/NIMO order was issued in connection with a generic proceeding in which the lost revenue issue was first subject to substantial public debate.

NIMO's bonus was 10% of savings. (There were numerous differences between the two utilities in the ways the savings were calculated and collected, so the shared-savings percentages are not on a comparable basis.)

During the proceeding commission staff argued that an incentive package that did not incorporate full ERAM-type decoupling of revenues from sales would work to the detriment of DSM. As a result, the commission directed O&R to work with staff on a decoupling proposal for later consideration.

A majority of the incentive mechanisms adopted since 1989 have emulated the shared-savings, or share of benefits, approach pioneered in the O&R/NIMO decision. Shared-savings bonuses appear to be finding favor with both utilities and regulators because:

• The concept is simple and readily understood by all parties, including the general public.

• They motivate both cost-effectiveness and greater spending on DSM. The utility can maximize its bonus by pursuing *all* opportunities for which benefits exceed costs.

• They are being formulated to ensure that a majority, usually 75% or more, of the net benefits of DSM accrue to ratepayers, thus minimizing the possibility that utilities will receive windfall profits.

Although simplicity is one of shared-savings mechanisms' virtues, greater complexity in the details of the mechanisms is a recent phenomenon. Bonus formulas adopted in Oregon for Portland General Electric, in Maine for Central Maine Power, and in Iowa in generic rules are basically shared-savings mechanisms, but the bonus cannot be described as a simple share of the benefits. Formulas are instead used to make the utility's share vary with cost-effectiveness, performance against a baseline, and/or performance in relation to the previous year's efforts. Whether this increasing complexity strengthens or weakens the motivating power of the mechanisms remains to be seen.

New England Electric System Incentives

In September 1989 New England Electric System, a Massachusetts-based utility holding company, submitted identical DSM incentive proposals in the three states where it operates retail subsidiaries: Massachusetts, New Hampshire, and Rhode Island. Like O&R and NIMO,

NEES proposed to receive as a bonus a portion of the savings resulting from its DSM programs. NEES did not, however, request any lost revenue adjustment, because its retail subsidiaries purchase electricity under an arrangement that tends to minimize lost revenues.[2]

By mid-1990, incentive mechanisms were approved in all three states, making NEES the first utility to pursue and receive incentives in more than one jurisdiction (Rhode Island PSC 1989; Massachusetts DPU 1990; New Hampshire PSC 1990). The mechanisms differ in the three states. The Massachusetts commission, in particular, thoroughly revised NEES's proposal for calculating the incentive, offering a fixed dollar amount per kilowatt and kilowatt-hour saved in lieu of a share of the savings. The company's success in multiple jurisdictions led several other New England utilities to formulate incentive proposals.

NEES's success in securing incentives is partly attributable to a close working relationship with the Conservation Law Foundation of New England (CLF), a Boston-based environmental advocacy group. CLF had originated the concept of the "collaborative process" for DSM program planning and design: a cooperative effort among the utility and other parties, who have traditionally intervened in rate cases, to reach consensus on DSM programs outside a formal commission proceeding. The results of the process are submitted to regulators for review and approval. Having played a significant role in the development of NEES's expanding DSM programs, CLF was willing to work with the company on the design of an incentive mechanism and to support the request for incentives in testimony before the three state commissions. Variations on CLF's collaborative process subsequently became the vehicles for considering incentives in several other states.

NEES's top management, led by CEO John Rowe, aggressively touted the company's success in achieving DSM incentives and thus gave the incentives movement new credibility, especially within the utility industry. In explaining his company's expansion of DSM budgets during a recession, Rowe said that "with current recovery and a fair opportunity for profit, our states have made [DSM] a real business." NEES executives also pointed out that the company allocated a greater portion of its budget to DSM than any other investor-owned utility in the country—in 1991, about 5% of revenue (ADSMP 1991). The inference was clear: incentives were beginning to fulfill the promise of expanding DSM.

---

[2] NEES's retail companies purchase power from an affiliated wholesale generating company. The wholesale rates are set annually by the Federal Energy Regulatory Commission (FERC). FERC uses a future test year that incorporates the expected effects of DSM.

The California Collaborative

The New England collaboratives were closely observed in California, among other states. In the early 1980s California was at the forefront of energy conservation efforts, but programs had faded due to lower energy prices and generating capacity surpluses. By 1989, the Public Utilities Commission concluded that the shrinking capacity surplus, air quality concerns, and improvements in energy efficiency technologies demanded another look at the role of DSM. When the Natural Resources Defense Council (NRDC) and other stakeholders suggested a statewide collaborative on DSM, the commission endorsed the concept and requested recommendations within six months. Fifteen parties, including the state's four major utilities, comprised the resulting collaborative.

The collaborative's January 1990 report, *An Energy Efficiency Blueprint for California*, devoted considerable attention to DSM incentives and outlined different pilot "shareholder incentive" (DSM bonus) proposals for each of the participating utilities (California Collaborative Process 1990). Owing to the number and prominence of the parties involved, especially Pacific Gas & Electric, the nation's largest investor-owned utility, release of the *Blueprint* was widely covered in the trade press. The visibility of DSM incentives increased, and "collaborative process" became an established term in the utility industry.

The collaborative's formulation of DSM bonus proposals represented an admission that ERAM was not enough: while ERAM removed the loss-of-sales penalty for DSM, ERAM alone was insufficient to overcome utilities' hesitancy. The utilities' incentive proposals that developed from the collaborative effort were approved by the commission in August 1990.

Two of the three electric utilities' mechanisms followed the shared-savings approach; the third was based on ratebasing the utility's investment. Each utility was also exposed to a financial penalty if it failed to meet program-specific performance objectives. Subsequent to the California collaborative the potential for penalties has appeared in many other incentive designs.

Penalties are unpopular with utilities, which claim that they introduce an element of risk that is counter to the spirit of providing DSM incentives in the first place. Proponents of penalties generally cite a desire for "symmetry" in mechanisms—if a bonus is extended for good performance, the reasoning goes, it is only fair to ratepayers that a penalty be levied for poor performance. A middle ground taken by some utilities and regulators is that penalties are inappropriate in the start-up phase of large-scale DSM programs but may be reasonable later on.

Like bonuses, penalty clauses are increasingly complex. To date, no utility operating under a DSM incentive mechanism has had a penalty levied for poor performance[3]—which could mean that penalty clauses are having the intended effect (i.e., utilities subject to penalties are motivated to achieve at least adequate performance) and/or that the level of performance needed to avoid a penalty is not very demanding.

The 1990 Orange and Rockland Incentive Revision

A fourth precedent-setting incentive action in mid-1990 again involved Orange and Rockland Utilities. By the spring of 1990 the New York commission staff was concerned that the incentives package established the prior September was not having the hoped-for effect on O&R's DSM programs. Company DSM plans filed with the New York Power Pool suggested that the reforms had produced few changes. O&R's plan emphasized load management, whereas the staff wanted to emphasize efficiency (conservation) programs. A general rate case in progress gave the commission an opening to revisit O&R's incentive mechanism and to consider the decoupling proposal that had resulted from the September 1989 order.

In August 1990 the commission approved a settlement in the case that instituted the "Revenue Decoupling Mechanism" (RDM) for O&R. Modeled on California's ERAM, it was the first full decoupling arrangement adopted in another state. Chapter 7 explains how the ERAM concept was adapted to New York's regulatory framework.

A revision of O&R's DSM bonus was teamed with the RDM. In lieu of the shared-savings approach implemented previously, the commission substituted an annual adjustment to the company's overall return on equity (ROE). The adjustment, which can be positive or negative, is a function of O&R's performance in two areas: cumulative energy savings, in kilowatt-hours, and net resource savings, in dollars. Whereas previously O&R's incentive depended only on net resource savings, the change added an incentive to emphasize kilowatt-hour reductions. Chapter 7 covers the details of the new bonus arrangement.

From a policy standpoint, the revisions to O&R's incentive mechanism highlight two points. First, specific utilities' incentive plans can be tailored to address perceived deficiencies, such as O&R's tendency under its initial incentive to emphasize demand savings over energy savings. Second, for the near term, incentive mechanisms will likely remain fluid. Commissions' prerogative to reconsider previous deci-

---

[3] Some utilities have had their authorized rates of return lowered in rate cases due to regulators' assessments that their DSM programs are deficient; such judgmental adjustments are outside the scope of incentive/penalty mechanisms considered here.

sions, especially in an area as experimental as DSM incentives, means that mechanisms established in a few years may be quite different.

In terms of mechanism design, the O&R revision heralded greater interest in ERAM and its variants. Subsequent to the O&R order, Washington State and Maine adopted decoupling for specific utilities, and a 1991 Connecticut statute authorized the commission to implement decoupling. Chapters 4 and 7 discuss recent implementations of the ERAM concept outside California.

One reason for the growing interest in ERAM is the belief that it may overcome utilities' sales promotion bias. In this view, utilities operating under DSM-specific lost revenue adjustments will not wholeheartedly embrace DSM, because incremental sales will still produce additional contributions to base revenue and profits.

A second factor is interest among some utilities in the revenue-stabilizing benefits of ERAM. While many utilities have vigorously opposed suggestions that ERAM be considered in their states, others are apparently shifting to the view that the loss in upside revenue potential implied by ERAM is an acceptable tradeoff for its risk-reducing benefits.

Sharp divisions on the merits of ERAM are not confined to utilities; consumer advocates, for example, can be found in pro- and anti-ERAM camps. Considering both these differences and the broad scope of the regulatory changes involved, ERAM will undoubtedly be a major point of contention for some time.

National DSM Incentive Developments

Interest in DSM incentives has expanded well beyond the states mentioned. Table 2-1 summarizes the status of incentive proposals, proceedings, and implemented mechanisms as of late 1991. The reader is cautioned that the information presented is a "snapshot" at a particular point in time and is subject to frequent changes.

Table 2-1. State-by-State Summaries of Recent DSM Incentive Developments

State	DSM Incentive Developments
Arizona	An October 1991 order allows utilities to recover expenditures on DSM programs preapproved by the commission staff. The commission also indicated it would consider allowing lost revenue recovery and bonuses.
California	Electric Revenue Adjustment Mechanism (ERAM), which adjusts for lost revenue, in operation since 1982. DSM shareholder incentive mechanisms developed collaboratively were approved by commission in August 1990 for San Diego Gas & Electric (SDG&E), Pacific Gas & Electric (PG&E), Southern California Edison (SCE), and Southern California Gas (SCG). SDG&E and PG&E mechanisms provide shared-savings bonuses for resource-oriented DSM programs. SCE received ratebase treatment for resource-oriented DSM programs. SCG receives a percentage of expenditures. All utilities are allowed markups on expenditures for certain low-income and informational programs and are subject to penalties for poor performance. A current rulemaking is considering revisions to incentives.
Colorado	Ratebasing of DSM expenditures was approved in a settlement with Public Service of Colorado (PSCO) in November 1990. PSCO will also collect an incentive on DSM provided through a bidding program equal to 5% of the estimated cost of a benchmark purchased power agreement. The incentive can be adjusted up or down according to a formula based on the life of the DSM and its cost. A docket was opened on decoupling in July 1991, and the issue is being addressed in a collaborative process.
Connecticut	A 1988 statute authorized DSM in rate base with a bonus return of 1 to 5%. Ratebasing, a variable bonus return of 1 to 3% based on program cost-effectiveness, and a partial sales adjustment mechanism were implemented for United Illuminating in a February 1990 order. A June 1991 statute authorized the commission to implement decoupling of revenues from profits.
District of Columbia	Ratebased recovery of program costs over 10 years was authorized for Potomac Electric Power Company in July 1990. An October 1991 order authorized lost revenue recovery and a variable shared-savings bonus.
Florida	DSM expenditures are recoverable through a balancing account. Capital-type DSM expenditures can be amortized with a return through the balancing account. A current docket is considering DSM incentives.
Georgia	A 1991 statute directs the commission to "consider lost revenues, if any, changed risks, and an equitable sharing of benefits between the utility and its customers" when establishing rates to cover the costs of approved DSM resources. The commission has deferred Georgia Power's proposal for a rider to recover DSM program costs and lost revenues, and a shared-savings bonus.
Hawaii	DSM incentives have been discussed in a docket on integrated resource planning, established in 1990.

Table 2-1 (continued)	
State	**DSM Incentive Developments**
Idaho	Ratebasing of DSM expenditures is allowed. A proceeding was opened in 1989 on DSM-related revenue loss. Utilities were directed to study the issue and propose mechanisms to adjust for lost revenue, if appropriate. Pacific Power & Light's proposal to sell energy efficiency services to commercial customers in return for a share of the customers' bill savings was approved in 1990.
Illinois	Illinois has approved special riders for recovering costs and lost revenues associated with pilot DSM programs, but not for full-scale programs.
Indiana	In October 1990 the commission approved a stipulation agreement providing deferred recovery of DSM program costs, lost revenues, and a shared-savings bonus of 10 to 20% (20% if its 1995 peak demand reduction target is fully achieved).
Iowa	A 1990 statute authorizes ratebasing of DSM; recovery of DSM expenditures outside general rate cases; and adjustment up or down in cost recovery based on DSM performance. Rules adopted in April 1991 allow lost revenue recovery and DSM bonuses of up to 25% of net benefits, as measured by the societal test. A penalty of up to 15% of planned net benefits could be assessed for poor performance.
Maine	Commission rules allow for either ratebasing or balancing account recovery of DSM expenditures. The commission has statutory authority to reward or penalize a utility up to 10% of DSM program costs based on performance. A DSM incentives proposal including an ERAM-type lost revenue adjustment and a shared-savings bonus was adopted for Central Maine Power in May 1991.
Maryland	A cost recovery and incentive mechanism negotiated in a collaborative process for Potomac Electric Power Co. (PEPCO) was approved in August 1991. It provides 5-year amortized cost recovery, lost revenue adjustment, and collection through a demand-side rider with annual true-up. The rider would be used only if PEPCO's rate of return falls below the authorized level. Program costs could be deferred until a rider is applicable. A bonus of 5% of savings could be obtained if performance exceeds program goals by 10%. Incentives for other utilities are being considered in separate collaboratives.
Massachusetts	DSM bonuses providing specific dollar amounts per kW and kWh saved were approved in 1990 for Massachusetts Electric Co. and Western Massachusetts Electric Co. (WMECO). Utilities collect program costs through fuel clauses or surcharges with balancing-account treatment. WMECO is authorized to recover estimated lost revenues.

Table 2-1 (continued)

State	DSM Incentive Developments
Michigan	Current conservation programs of major utilities are funded through a per-kWh surcharge, with reconciliation of revenues and expenditures. An incentive mechanism providing a cents/kWh bonus on a sliding scale of cost-effectiveness is in place for certain conservation programs of Consumers Power. A May 1991 order regarding Consumers Power established a minimum DSM spending goal of 2.5% of revenue requirements; authorized 10-year ratebasing of DSM expenditures; said lost revenue recovery will be considered in a reconciliation proceeding after two years; and indicated the commission may apply an ROE adjustment of −2% to +1% depending on actual DSM spending and cost-effectiveness.
Minnesota	A March 1991 order approved a Northern States Power proposal to rate-base certain conservation program expenditures, earn a bonus return of up to 5% on the unamortized balance, and recover 50% of revenue lost through promoting its interruptible service tariff.
Nevada	Utilities can earn an AFUDC-type return on DSM expenditures between rate cases. Amortized cost recovery over 3 to 5 years without a return is permitted. A proposed rule issued in September 1991 would allow recovery of DSM-related lost revenues and a bonus of 10% of net savings.
New Hampshire	Balancing-account recovery for DSM expenditures was approved for Granite State Electric in January 1990. An August 1990 order approved in concept the recovery of DSM-related lost revenues, and also approved a shared-savings bonus for Granite State Electric providing 5% of gross benefits and 10% of net benefits.
New Jersey	Regulations adopted in October 1991 allow utilities to earn DSM incentives based on a shared-savings approach and/or a "standard offer" under which the utilities could compete with other providers to furnish DSM services at a fixed price. Utilities can propose a method to recover lost revenues.
New York	Some DSM program costs are recovered through fuel adjustment clauses; others are recovered in base rates. Incentive mechanisms approved in 1989 and 1990 for 7 utilities provide lost revenue adjustments and bonuses of 5% to 20% of net savings from DSM. The incentive mechanism for Orange and Rockland Utilities was changed in August 1990 to include revenue decoupling (similar to California's ERAM) and a bonus or penalty on the company's overall rate of return tied to DSM performance. The mechanism for Consolidated Edison was revised in April 1991 to follow the Orange and Rockland model.
North Carolina	A May 1990 order said utilities could initiate deferred acounting, with a return, for DSM expenditures. Four utilities filed DSM cost recovery/incentive proposals in May 1991; commission action is pending.

Table 2-1 (continued)

State	DSM Incentive Developments
Ohio	An April 1991 order in a generic proceeding approved the commission staff's proposed policy to allow deferred recovery of non-test-year DSM expenditures, with carrying charges; recovery of lost revenues; and a shared-savings bonus of 10% of net savings.
Oregon	Pacific Power & Light was authorized in June 1990 to provide energy efficiency services to commercial customers in return for a share of the customers' bill savings. Portland General Electric was authorized in January 1991 to recover DSM-related lost revenue and a shared-savings bonus. A docket was opened in July 1991 to consider policies, including incentives, to encourage acquisition of cost-effective DSM.
Pennsylvania	Utilities submitted cost recovery and incentive proposals in January 1991, along with DSM plans. Commission staff has proposed recovering program costs and lost revenues through a surcharge, and performance-based bonuses. A current docket is considering this proposal.
Rhode Island	Newport Electric was authorized in 1989 to capitalize DSM expenditures and recover them over 5 years through its purchased power adjustment clause. Narragansett Electric's request for balancing-account recovery of DSM expenditures and a shared-savings bonus was approved in 1990. The bonus is equal to 5% of gross benefits and 10% of net benefits above a threshold of approximately 50% of program goals.
Vermont	DSM expenditures can be ratebased or expensed. Recovery of DSM expenditures not already in rates and lost revenues was authorized by an April 1990 order establishing ACE (Account Correcting for Efficiency) mechanism. A January 1991 order established a framework for a shared-savings incentive and directed parties to develop a mechanism consistent with the framework for Green Mountain Power (GMP). As the result of a settlement with a consumer group, GMP subsequently submitted a DSM plan but did not request an incentive.
Virginia	A generic proceeding was initiated in January 1991 on DSM cost recovery and incentives. An April 1991 staff report recommended that utilities be directed to file proposals for program cost recovery and treatment of lost revenues, and recommended that consideration of incentives be deferred to a later proceeding. Commission action is pending.
Washington	A 1980 statute allows certain DSM expenditures in rate base with a 2% bonus return on the equity portion. A mechanism approved for Puget Sound Power & Light in April 1991 includes a rate adjustment mechanism to provide current recovery of DSM and purchased power costs and a per-customer revenue adjustment (decoupling) mechanism. Puget's proposal for a DSM performance bonus is pending.

Table 2-1 (continued)	
State	**DSM Incentive Developments**
Wisconsin	The commission has instituted several different performance-related bonus mechanisms. From 1987 to 1990, Wisconsin Electric Power was eligible for a 1% bonus on the equity portion of ratebased DSM for each 125 MW of realized savings. In January 1991 the commission rejected WEPCO's request for a 10% share-savings bonus; instead, it authorized $500,000 to be used for staff bonuses related to DSM in 1991.

References

Association of Demand-Side Management Professionals (ADSMP). 1991. "NEES Credits Regulatory Incentives in 'Overwhelming' 1990 DSM Success." *Strategies* 2 Spring.

Barakat, Howard & Chamberlin, Inc. 1988. *Status of Least-Cost Planning in the United States*. Palo Alto, Calif: Electric Power Research Institute.

Blackmon, Glenn. 1991. "Conservation Incentives: Evaluating the Washington State Experience." *Public Utilities Fortnightly* 127. January 15.

California Collaborative Process. 1990. *An Energy Efficiency Blueprint for California*. January.

Energy Conservation Committee of the National Association of Regulatory Utility Commissioners. 1988. "Statement of Position of the NARUC Energy Conservation Committee on Least-Cost Planning Profitability." July 26.

Massachusetts Department of Public Utilities. 1990. Order 89–194/195. March 30.

Moskovitz, David. 1988. "Will Least-Cost Planning Work Without Significant Regulatory Reform?" Presented to NARUC Least-Cost Planning Conference, Aspen, Colo. April 12.

New Hampshire Public Service Commission. 1990. Order No. 19,905. August 7.

New York Public Service Commission. 1988. Case 29409, Opinion No. 88-20. July 26.

Rhode Island Public Service Commission. 1989. Order in Docket No. 1939. December 27.

Rowe, John. 1990. "Making Conservation Pay: The NEES Experience." *The Electricity Journal* 3. December.

Washington. 1980. Rev. Code Wash. 80.28.025.

Wisconsin Public Service Commission. 1986. Order in Docket No. 6630-UR-100. December 30.

Chapter 3

California's ERAM Experience

Chris Marnay and G. Alan Comnes

Introduction

In 1982, the California Public Utilities Commission (CPUC) introduced a new regulatory procedure called the Electric Revenue Adjustment Mechanism (ERAM). ERAM periodically adjusts the non-fuel part of rates, known as base rates, to ensure that an electric utility collects its full authorized revenue requirement, despite its level of sales. ERAM achieves this parity by maintaining an account that tracks revenue miscollections. The balance in this account is amortized in future rates.

ERAM was adopted primarily for two reasons. First, it was accepted in a period of turmoil for California's electric utilities in the hope that eliminating sales forecasting error would reduce utility risk and improve the financial health of the industry. Second, ERAM was intended to remove a perceived anti-conservation bias of previous California regulation. As described in Chapter 1, this bias derives from an economic incentive for electric utilities to market power between rate cases, providing a significant impediment to adopting cost-effective, demand-side resources.

ERAM enjoys wide support in the utility industry in California, is enthusiastically endorsed by conservationists (Cavanagh 1988), and is currently in effect for four of the five largest investor-owned utilities.[1] Most California utilities support ERAM (Woo and Peters 1989), and

---

[1] The five largest electric utilities regulated by the CPUC are Pacific Gas and Electric (PG&E), Southern California Edison (Edison), San Diego Gas and Electric (SDG&E), Sierra Pacific Power (SPP), and Pacific Power and Light (PP&L). ERAM has been adopted for all of these utilities at some time but, for reasons discussed in this chapter, has been eliminated for PP&L.

the National Association of Regulatory Utility Commissioners' Energy Conservation Committee has a record of supporting ERAM-like rate-making reforms (NARUC 1988). However, the CPUC staff is divided; some staff members have recommended eliminating ERAM. Independent analysts also have reservations (Sissine 1989, Murray 1991, Reid and Weaver 1991).

Since California adopted ERAM, the introduction of similar revenue decoupling mechanisms (RDMs) has been considered by several states; the number of states that have adopted RDMs is growing.

Mechanics of ERAM

ERAM is conceptually simple: utilities are allowed to recover only authorized levels of non-fuel-related revenue requirements. Such revenues are known in California as base-rate revenues. The revenues that appear on financial statements are authorized amounts rather than actual amounts. Any difference between the two is tracked in a balancing account. The ERAM account earns interest to compensate either shareholders or ratepayers for the time value of money, thus removing any incentive for the utility to build large negative balances in the account as a way to obtain cost-free capital.[2]

An adder to retail rates called the ERAM rate is usually set once a year to amortize the ERAM balance over a 12-month period. The ERAM rate can be positive or negative, depending on the sign of the ERAM account balance. This rate is added to base rates for the current period to form the *effective* base rate. This rate plus the cost of fuel is the actual retail rate paid by customers.

Simplified ERAM Example

A simplified example of how ERAM works appears in Table 3-1. A more complete example is provided in Marnay and Comnes (1990). Table 3-1 shows ERAM's operations for a hypothetical utility over a two-year period. The table is divided into two parts. Section I shows key parameters set by the ratemaking process. Section II shows actual results without ERAM (subsection A) and with ERAM (subsection B). The example was developed using the following assumptions:

- Only base-rate revenues are shown. Authorized base-rate revenues are assumed to be $4,250 million per year, including $3,500 million

---

[2] The ERAM account accrues interest at rates competitive with commercial paper. Utilities do not earn authorized rates of return on the ERAM balance because, unlike money invested in actual operations, the ERAM balance is a low-risk investment.

in non-fuel operating costs and $750 million in authorized earnings. These authorized revenues are shown on lines 1 through 4. The example assumes that authorized revenues are the same in both years. In practice, rate cases or attrition proceedings change authorized revenues on a yearly basis.

- Fuel costs, which are covered by a separate fuel adjustment clause in California, are ignored.

- The effects of capital structure and corporate income taxes are ignored.

- Before the beginning of year 1, the utility's sales are forecasted to be 63,400 GWh. During year 1, a sales forecast of 68,640 GWh is made for year 2. These forecasts appear on line 6.

- Actual sales are higher than forecast in year 1 and lower than forecast in year 2. Actual sales are shown on line 10.

- Actual operating costs equal forecasted costs in year 1 but are lower than forecast in year 2. Actual costs are shown on line 11.

- Flat rates (i.e., no demand charges or tiered rates) and one customer class are assumed. Thus, base rates are set by simply dividing authorized revenues by forecasted sales, as shown on line 7.

- Each year a rate is determined to amortize the ERAM account based on the latest sales forecast, as shown on line 8.

- Actual revenues in each year equal actual sales times the tariffed rate. Revenues collected are shown on line 12 in the without-ERAM case and on line 15 in the with-ERAM case.

- Costs are assumed to be invariant with short-term changes in sales. Because the rates shown ignore fuel costs, this is a reasonable assumption. In year 2, the example assumes that costs are lower due to cost-cutting measures taken by management.

- The ERAM balance at the beginning of year 1 is assumed to be zero, as shown on line 5. Funds in the balance are assumed to earn 8%.

The example shown in Table 3-1 is designed to demonstrate two main points. First, with ERAM, earnings are driven by *authorized* revenues, not the actual revenues that are collected by the company. Second, both with and without ERAM, utility shareholders are affected by changes in operating costs.

**Table 3-1. Electric Revenue Adjustment Mechanism
Hypothetical Example for a Two-Year Period
($million, unless noted otherwise)**

	Year 1	Year 2	Formula
I. RATES AS SET BEFORE THE YEAR BEGINS			
Forecast (authorized) base-rate (non-fuel) revenues:			
(1) Operating costs	3,500	3,500	
(2) Authorized rate of return	12.5%	12.5%	
(3) Rate base	6,000	6,000	
(4) Authorized base-rate revenues	4,250	4,250	$(1)+[(2)\times(3)]$
(5) ERAM balance	0	(181)	(20) of previous year
(6) Forecasted sales (GWh)	63,400	68,640	
(7) Base rate (cents/kWh)	6.70	6.19	$[(4)/(6)]\times100$
(8) ERAM rate (cents/kWh)	0.00	-0.26	$[(5)/(6)]\times100$
(9) Effective base rate (cents/kWh)	6.70	5.93	$(7)+(8)$
II. ACTUAL EVENTS DURING THE YEAR			
(10) Sales (GWh)	66,000 (higher than forecasted)	65,900 (lower than forecasted)	
(11) Operating costs	3,500 (same as forecasted)	3,473 (lower than forecasted)	
A. Impact Without ERAM			
(12) Revenues collected and reported	4,424	4,080	$[(7)\times(10)]/100$
(13) Earnings reported	924	607	$(12)-(11)$
(14) Rate of return	15.4%	10.1%	$(13)/(3)$
B. Impact with ERAM			
(15) Revenues collected	4,424	3,906	$[(9)\times(10)]/100$
(16) Revenue deviation	(174) (over-collection)	344 (under-collection)	$(4)-(15)$

Table 3-1 (continued)

| | Year | | |
	1	2	Formula
II. B. Impact With ERAM (continued)			
(17) Revenues reported	4,250	4,250	(15)+(16) [equal to (4)]
(18) Earnings reported	750	777	(17)−(11)
(19) Rate of return	12.5%	13.0%	(18)/(3)
(20) End-of-year ERAM balance	(181)	162	(5)+(16)+ 8% interest on average balance

Results Without ERAM

Sales are higher than forecasted in year 1 and are lower than forecasted in year 2. Financial impacts in the without-ERAM case are shown in part II.A of Table 3-1. Without ERAM, actual sales significantly affect earnings. Line 14 shows that earnings are above the authorized 12.5% level in year 1 (15.4%) and below the authorized (10.1%) in year 2. Year 2 earnings would have been even less (9.67%) without the cost-cutting measures taken by the utility.

Results with ERAM

ERAM results are shown in part II.B of Table 3-1. In year 1, sales are higher than forecasted, but the additional revenues are credited to the ERAM account (line 16) rather than reported on the utility's income statement. Since reported revenues are precisely those authorized (line 17), the rate of return in year 1 is exactly as authorized: 12.5%. The balance in the ERAM account, including interest, is carried over from year 1 (line 20) to year 2 (line 5) and is returned to ratepayers in year 2 via an ERAM rate credit (line 8).

In year 2, sales are lower than forecasted. Costs are also lower due to productivity improvements implemented by the utility. With ERAM, the utility is insulated from the effects of the lower sales and keeps the benefits from improving productivity. Thus, the rate of return in year 2 with ERAM is 13.0%, or 0.5% above the authorized level.

Context of California Regulation

While ERAM itself is simple, it does not operate in isolation; a full appreciation of its impact requires a general understanding of electric ratemaking in California. The following describes how the CPUC sets the fuel and base components of rates for its major electric utilities.

General Rate Cases

California regulation deviates from the national norm in that, since 1984, general rate cases (GRCs) for large electric and natural gas utilities are usually conducted at regular three-year intervals, rather than whenever the utility or Commission staff files for a change in rates. GRCs in California focus only on the non-fuel revenue requirement of the utility—that is, recovery of depreciation, return on investment, taxes, non-fuel-related operation and maintenance expenses, and administrative and general expenses. Fuel costs, considered to be more volatile over time, receive periodic rate adjustments via the Energy Cost Adjustment Clause (ECAC) proceeding. The non-fuel revenue requirement is known as the utility's *base-rate revenue requirement*, and the rate component that emerges from the GRC is called the *base rate*. All GRC and ECAC calculations are based on a future test year. In contrast, most commissions set rates using historical test years (Phillips 1988).

Attrition

Attrition allows for adjustments in the base-rate revenue requirement of the utility in years that are not covered by GRCs. Attrition attempts to measure several specific factors that will make authorized revenue requirements inappropriate over time, such as inflation, productivity change, customer growth, and fluctuations in the cost of capital. Many states have introduced attrition mechanisms, although implementations differ (Radford 1988).

In California, in each year that a GRC is not held, utilities file for an attrition revenue adjustment. Attrition was originally intended to allow mechanical, noncontroversial adjustments to the base-rate revenue requirement using a methodology agreed on in the previous GRC or in a general policy-setting proceeding. On several occasions, however, the CPUC has used attrition proceedings to authorize revenues for utility programs not considered in the GRC. Thus, attrition proceedings have sometimes resembled mini–rate cases rather than mechanical, simplified proceedings.

In California attrition includes operational and financial components. *Operational attrition* adjusts for the utility's cost of service in years between rate cases. Operational attrition is generally divided into expense and capital. The expense portion of attrition allows for inflation on certain expenses and considers the impacts of productivity and customer growth on base-rate expenses. The capital portion of attrition, sometimes called rate-base attrition, accounts for changes in the rate base resulting from forecasted plant additions.

Financial attrition[3] compares a utility's debt and equity returns to changing financial markets. The original intention of financial attrition was to compensate the utility for exogenous changes in financial markets, particularly interest rates. The proceeding has recently been expanded to include a review of the equity returns of utilities, and financial attrition hearings are now held annually. This annual realignment of the rate of return has arguably produced a further reduction of financial risk to electric and gas utilities in California.

Authorized base-rate revenues change at least once a year as a result of GRCs or attrition proceedings. Rates are always set using the latest available sales forecast, typically the one generated in the most recent ECAC proceeding. Because changes to base-rate and ECAC revenue requirements occur at least once a year and not necessarily at the same time, tariffed rates change at least once a year.

ERAM Linkage to Attrition

There is an important link between ERAM and attrition. In states without attrition, the ratemaking process assumes an implicit relationship between sales and costs. In such states, a utility's costs may increase between GRCs, but the utility has a fair opportunity to earn its authorized rate of return because increased revenues from increased sales roughly cover the additional costs. This assumed relationship helps explain why states that have no attrition and use historical test years have comfortably set rates for many years, although the test year data do not produce accurate estimates of a utility's future costs.

ERAM breaks the implicit linkage between sales and per-unit base-rate costs. Because the test year revenue requirements are accurate only for the test year, this decoupling challenges the accuracy of the authorized base-rate revenue requirement in the years that follow the test year. Attrition partially *recouples* revenues to costs by adjusting the revenue requirement for estimated changes in costs associated with inflation, productivity, and customer growth.

---

[3] Financial attrition is also known as the annual cost of capital proceeding.

ECAC/AER

Fuel-related costs, including payment to nonutility generators (NUGs), are recovered in rates set in the Energy Cost Adjustment Clause (ECAC) proceeding (Ameer 1989).[4] In the ECAC, differences between authorized and actual fuel costs are tallied in a balancing account and amortized in future rates. During much of the 1980s, a fraction of the fuel budget was withheld from the ECAC balancing account and collected in an Annual Energy Rate (AER). The AER is a fixed fraction of fuel costs for which the utility, rather than the rate-payer, is liable. The AER provides an incentive for utility management to minimize fuel costs. However, because of the difficulty in forecasting fuel costs, the AER was often inaccurate and the CPUC frequently suspended it, and in August 1990, the CPUC suspended the AER indefinitely. As an incentive mechanism, the AER clearly failed.

Utilities may file ECAC filings as often as twice a year, although the typical frequency is once per year. Sales forecasts made as part of the ECAC proceedings provide a convenient source for determining the ERAM rate.

Development of ERAM

Origin of ERAM

The idea of decoupling revenues from sales has its origins in the regulation of the natural gas industry. Since 1978, California has balanced gas utility revenues to make utilities indifferent to the actual level of sales relative to forecast levels (Barkovich 1989). RDMs were adopted for California gas utilities partly to promote conservation but mostly to reduce the utility revenue variability that was caused by weather-sensitive customer demands along with the natural gas supply curtailments of the late 1970s.[5]

In electricity, the CPUC first considered a decoupling mechanism as part of a landmark rate case for PG&E in late 1981 (CPUC 1981). Costly delays in nuclear construction had left California's utilities in financial difficulty and the state faced an imminent generating capacity shortfall. PG&E pleaded for significant rate relief and for the adoption of various regulatory reforms that would ease the company's debt bur-

---

[4] In other states, ECAC-like mechanisms are commonly known as fuel adjustment clauses, fuel cost adjustments, or fuel offset mechanisms.

[5] Since 1988, RDMs have been partially eliminated for large natural gas customers as a way to give natural gas utilities an incentive to lower transportation costs and to maximize system sales.

den and raise its rate of return. The prime interest rate stood at 19%; avoiding borrowing for capacity additions was the paramount goal. In this economic climate, eliminating disincentives to conservation was imperative, as well as stimulating development of independent sources of generation. In its rate case decision, the Commission did two things to help PG&E's financial health: it granted a higher rate of return and it adopted ERAM (CPUC 1981). The language of the decision indicates that ERAM was adopted as much for financial reasons as it was to promote conservation.

PG&E had proposed ERAM in its rate case filing, and its proposal received support by the CPUC staff and the California Energy Commission staff, although their proposed mechanisms differed slightly. ERAM was later adopted for the two other major California investor-owned utilities (IOUs), Edison and SDG&E, and finally, for SPP and PP&L. The five major IOUs together account for about 80% of electricity sales in the state.

Mid-1980s ERAM Review

Conditions in the industry looked quite different in 1985, when the CPUC initiated a seminal review of ratemaking in the state. The addition of three large nuclear plants partially or wholly owned by California utilities, and the unexpectedly rapid emergence of NUG capacity, had produced a comfortable level of capacity. Furthermore, fuel prices had fallen, and the competition created by the emerging NUGs made utilities more rate conscious. ERAM was being reviewed in a different environment from the one into which it was introduced.

A thorough analysis of the incentive structure created by ERAM formed a central part of the Commission's investigation (CPUC 1985, 1986). Commission staff member Mark Ziering wrote a comprehensive survey of the key issues, which was attached to the initial Order Instituting Rulemaking (OIR) issued by the Commission (Ziering 1986). According to Ziering the era of ERAM was over, and the pressing need of the time was preventing uneconomic bypass, which occurs when large customers generate their own power although they could be served less expensively by a utility at a price above its marginal production cost.[6] Customers would be best served, Ziering argued, if the state's IOUs were freer to make the favorable agreements with cus-

---

[6] Since fuel costs are passed through to ratepayers in the ECAC process, the marginal production cost that an IOU is at risk for is trivial. However, the CPUC had decided that no sales should be made below the average fuel cost of generation so that marginal production cost would be approximated by the average ECAC fuel cost.

tomers that bypass prevention demanded. Such a role for the utilities, however, is quite inconsistent with ERAM, as Ziering noted:

> The current ECAC and ERAM mechanisms, however, largely insulate utility earnings from changes in sales volumes. If utilities fail to take actions or to grant discounts where these are needed, or grant larger discounts than are required, there is no immediate effect on their earnings.

Ziering foresaw a strong incentive for the utilities to actively market their power if ERAM were eliminated. Removing ERAM would encourage the utilities to find customers for the electricity generated by their excess capacity, and this would benefit all ratepayers. All parties, like Ziering, emphasized the dangers of uneconomic bypass and the need to fully use the state's adequate supply resources; however, Ziering specifically pointed out the incompatibility of ERAM with these goals, placing its elimination firmly at the top of his list of priorities: "The most pressing [change] . . . to the current regulatory system is the elimination of ERAM and attrition mechanisms." The emphasis given to this proposal is noteworthy in a document that reviews virtually every aspect of ratemaking in the state.

Move to Eliminate ERAM

Partial ERAM Removal. The Commission, after a lengthy period of consideration, including a pivotal *en banc* hearing in March 1987, decided to partially adopt Ziering's recommendation that ERAM and attrition be eliminated (CPUC 1987). The Commission explicitly recognized that for a considerable period conservation efforts would be scaled back and, consequently, the need for ERAM diminished. Eliminating ERAM was consistently opposed by environmental groups, notably the Natural Resources Defense Council (NRDC), which questioned the conventional wisdom that the state's capacity glut would last for some time.

The Commission chose to distinguish the Large Light and Power customer class (LL&P) from other ratepayers, ruling that ERAM would be eliminated for LL&P but retained for other customer classes. This compromise, the Commission argued, would retain the correct conservation incentives for the latter classes of customers, while exposing and conditioning the utilities to competition among LL&P customers. The date set for the partial removal of ERAM was April 1, 1988. The separation of LL&P customers from the other rate classes was natural, given that most of the arguments in favor of ERAM elimination had focused on LL&P customers and that uneconomic industrial bypass was a primary concern at that time. Industrial customers were perceived as both the problem, since they posed the most credible bypass threat, and the potential solution, since their demand could quickly increase and exhaust the excess supply.

Implications of CPUC Decision. Despite the apparent simplicity of the CPUC's decision, it was actually radical. It proposed, in effect, to create a split utility that would act as a traditional utility subject to preexisting ratemaking, including ERAM, towards non-LL&P customers. Simultaneously, however, the utility was to act like a tough competitor towards its LL&P customers, fending off both the bypass threat and competition from other suppliers. The Commission essentially proposed institutionalized price discrimination. In the inelastic market segment—non-LL&P customers—total recovery of all authorized costs would be guaranteed by ECAC and ERAM. In the more elastic LL&P segment, the utility was to be allowed to discount to retain profit from incremental sales, on the theory that shareholders would have adequate incentive to maximize revenues without ERAM.

Many participants, particularly the utilities, felt implementing the Commission's decision would pose a great administrative burden. Ziering, in fact, had considered such a policy and predicted that it would result in ". . . the most complex system to date" (Ziering 1986).

Reversal on the Decision to Remove ERAM. Implementation was to begin with hearings scheduled for July 1988. The definition of the LL&P class became generalized to include all customers with peak demands over 1 MW, and the implementation date slipped into 1990. However, the hearings were never held, and one by one the interested parties began appealing to the Commission to reverse its ruling. There were behind-the-scenes negotiations on how ERAM should be reformed, if at all. The first formal evidence that these talks had led to a quiet Commission reversal came in a December 1988 decision (CPUC 1988). This decision reports a stipulation among parties active in the rulemaking process that essentially abandons the attempt to remove ERAM, reporting two key points to justify the change:

> First, it [the stipulation] concludes that "the likely level of any future uneconomic bypass can be dealt with under current procedures" without developing different treatment for a newly created less restricted class (LRC) of large customers.
> Second, the stipulation states that segregating the LRC for different treatment requires "a very complex ratemaking structure with potentially conflicting incentives," and the parties recommend that the

Commission not pursue its development of the separate LRC (CPUC 1988).

The first statement is supported by the absence of a significant rush of customers to generate their own power. This statement generally reflected the understanding that the Commission's move towards more cost-based ratemaking structures had not caused major disruption in the industry. The second statement supported the argument of ERAM advocates that partial elimination would produce contentious rate proceedings and create additional layers of regulatory accounting.

In later comments on the rulemaking proceeding filed with the Commission by the three major IOUs, their determination to keep ERAM is obvious. PG&E, for example, argued vehemently for a status quo approach:

> In PG&E's view, the current ratemaking mechanism [that includes ERAM] is a progressive approach to regulation that has been proven to be beneficial to ratepayers by providing utilities with incentives to keep rates down while offering innovative rates and demand-side management options. (Woo and Peters 1989)

In the end, the status quo prevailed.

ERAM Status Quo

The CPUC officially changed its position on eliminating ERAM in May 1989 (CPUC 1989). ERAM remains embedded in the CPUC's regulatory framework for major electric utilities. Ziering's recommendation to completely remove ERAM and attrition for the major companies is no longer under direct consideration, and no party has successfully argued that overcoming the practical barriers to implementing a partial ERAM is worthwhile. Also, memories have been short with regard to the financial benefits the first drafters of ERAM intended. While the state's utilities no longer need this support, they respond strongly to a threat of its removal. The conservation lobby has also been effective in its consistent support for ERAM.

Weak Opposition to ERAM. Two parties are potential losers as a result of ERAM: ratepayers, due to the transfer of sales risk from the utility in the form of more variability in rates; and NUGs, because they compete with utilities, for which ERAM provides a financial edge. In general, however, opposition to ERAM has been weak.

There are three possible reasons why opposition from small ratepayer groups is weak. First, the net effect on rates of ERAM is small and attracts little attention. Second, if ERAM does lower the utility's capital costs, this benefit may well be captured by ratepayers.[7] Third, ERAM removes the sales forecast as a major point of contention in rate cases; this may be seen as a significant benefit by small ratepayer groups, since their resources for participating in rate cases are limited.

The lack of opposition in California from large customer groups is perplexing, since industrial rates are relatively high (Marnay 1989), and industrial customers elsewhere, notably in New York, have bitterly opposed RDMs. The most plausible explanation is that large customers can avoid full tariffed rates. The Commission has offered a number of loopholes utilities can use to offer lower rates to large industrial customers, such as through individually negotiated contracts. A second loophole is adoption of an interruptible tariff. These tariffs are much cheaper than firm rates and the risk of interruptions is extremely low, due to the state's ample supply.

NUGs are losers since ERAM helps utilities financially, giving them a competitive edge; however, NUGs tend to have mixed goals. First, NUGs have generally received favorable treatment from the CPUC, so they are not directly competing with the utilities. Second, their longer-term goal to be free to compete with the IOUs must be tempered by a shorter-term need to maintain the utilities as financially healthy buyers of their output. Finally, the guaranteed market that NUGs enjoy gives them some advantage in raising capital (Pearl and Luftig 1991).

New Ratemaking Directions. In any event, the nature of the debate over ratemaking reform has changed since 1989. The main focus has shifted from ERAM towards renewed interest in DSM, including demand-side bidding and performance-based ratemaking (SCP 1990; see also Chapter 6). Most significantly, after the hot summer of 1988 and the permanent closure of the Rancho Seco nuclear generating station in 1989, the state was again approaching capacity shortfalls. Environmental issues, notably the state's infamous air quality, also received renewed attention. A determination to reverse the state's slide in DSM spending followed (CEC/CPUC 1988, Calwell and Cavanagh 1989, Messenger 1989). Additionally, the bypass threat has subsided, reducing the pressure for lower industrial tariffs.

Elimination of ERAM for PP&L. Ironically, about the same time that the decision to defer modifying ERAM for the state's major electric utilities was issued, the CPUC granted the request of Pacific Power

---

[7] No attempt has been made by the authors to quantify the impact, if any, of ERAM on authorized rates of return. Because of the many factors that affect utility risk, a quantitative estimation of ERAM's effect would be difficult.

and Light (PP&L) to eliminate ERAM entirely. PP&L argued that ERAM produces rate instability by causing frequent rate changes and reduces its incentive to offer competitive rates.[8] The Commission granted PP&L's request on the condition that PP&L not reduce its efforts to promote conservation in California. To this end, PP&L was ordered to file a plan to link earnings in California with its DSM efforts (CPUC 1990a). While PP&L has a small service territory within the state, the Commission's decision in this case shows that it will support the elimination of ERAM if there are other mechanisms to ensure utility commitment to conservation.

Developments in Other States

RDMs are being considered in several states and the number adopting them is growing. As of this writing, RDMs have been adopted in New York, Washington State, and Maine. After California, New York is farthest along in implementing decoupling mechanisms. Although New York established conservation-specific revenue adjustment mechanisms for its electric utilities in 1989 (Cole and Cummings 1990), it shifted to an RDM for Orange and Rockland Utilities (O&R) in 1990 (NYPSC 1990) and for Niagara Mohawk in 1991 (NYPSC 1991). The O&R mechanism is described in Chapter 7. A continuing generic investigation into incentives for conservation programs (NYPSC 1989), as well as rate reviews of individual utilities, may lead to the adoption of RDMs for other electric utilities in New York.

In contrast to California, RDMs face stiff opposition in New York from large industrial customers who argue that RDMs reduce a utility's incentive to minimize costs because it is no longer at risk for lost sales. Further, because RDMs have been combined with positive incentives for DSM, industrial customers believe they will be forced to subsidize programs that largely benefit customers in other classes (Murray 1991).

Effects of ERAM

Removes Conservation Disincentive

The conservation argument holds that without ERAM, a California utility encounters three incentives that may have adverse effects on conservation policy goals. First, between rate cases, the utility has an

---

[8] Unlike the larger electric utilities in California (PG&E, SCE, and SDG&E), PP&L is not subject to the general rate case and attrition plan. Thus, with the elimination of ERAM, PP&L would no longer be required to make annual changes to its base rates.

incentive to sell as much power as possible, as long as rates exceed the marginal cost of generation. The revenue gained from selling a kWh above the forecast level represents a direct contribution to the company's profits. Second, cost-cutting between rate cases benefits the utility because rates have been fixed assuming the higher costs. Conservation expenditures can be axed, like any other element in the company budget. Third, when costs of a conservation program have been added to customer rates, the utility benefits if the program fails to deliver the conservation promised. In this way the utility recovers the costs of the program yet avoids the revenue loss its success implies.

There is little doubt that ERAM does the job with respect to the first and third perverse incentives. Since under ERAM, the utility should be indifferent to its sales level, the utility is not punished for effective conservation programs, and its incentive to market power between rate cases is eliminated. It cannot be said, however, that ERAM provides a strong incentive to promote DSM when its primary effect is to remove disincentives against pursuing DSM. Positive incentives for DSM depend on favorable treatment of DSM investments.

Turning to the second perverse incentive, in California, conservation-related costs have been traditionally passed through as expenses and not given recovery as a rate base investment, although both accounting methods are used. Therefore, even with ERAM, the utility can benefit by reducing program spending, although the ability to do so is limited by prudence review. Further, as described in Chapter 6, California has recently adopted positive incentive mechanisms such as shared savings programs (SCP 1990, CPUC 1991).

Finally, discouraging utility marketing is not unambiguously good. Electrification of vehicles and industrial processes can offer real benefits, such as cleaner urban air (CEVTF 1989). Inhibiting interfuel competition by discouraging electric utility marketing without equivalent constraints on other energy companies could lead to an undesirable societal energy mix.

Impedes Deregulation

The most disturbing problem with ERAM is its effect on emerging competition in the industry. ERAM was introduced and survives using the economics of traditional regulated monopolies. Contemporary utilities, however, are not natural monopolies in all sectors of the electricity market but face stiff competition from bypass, independent power production, and even from independent sources of DSM (DeForest et al. 1990, Plummer and Troppmann 1990). Explicit or implicit in most discussions of electric power industry deregulation is the assumption that if fair rules regarding the entry of independent power producers and rules regarding transmission access are established, then a viable competitive market will be created. Commentators on deregulation have generally failed to recognize the importance of the existence of ERAM; even writers describing the California industry give it minor consideration (Plummer and Troppman 1990, Friedman 1991, Gilbert 1991). In fact, ERAM creates an uneven playing field at the outset, because a company that is protected from the effect of sales variations simply is not competing fairly in a free market.

Two aspects of this imbalance merit special attention. First, under ERAM, a utility has a guaranteed revenue stream; it is therefore a lower risk investment than a directly competing NUG and enjoys an unfairly low cost of capital. Second, the company has no incentive to be a tough negotiator with bypassers or to be tough in other contract negotiations affecting sales, to the detriment of ratepayers.

The latter problem was exemplified by the special contracts policy of the CPUC (Marnay 1989). Beginning with its proposed rulemaking to eliminate ERAM, the Commission clearly indicated its interest in negotiated rates to reduce uneconomic bypass. The utilities, in turn, actively pursued special contracts. PG&E, for example, negotiated more than ten special contracts between 1986 and 1988. The utilities' motives for negotiating the contracts were more strategic than financial, as ERAM was still in force.

When it became apparent in 1989 that ERAM would not be eliminated, the CPUC reviewed the special contracts in some contentious proceedings. It found several of PG&E's contracts unreasonable because they did not ensure that prices were above marginal cost (CPUC 1990b). This hindsight review, necessitated by ERAM, significantly reduced utilities' interest in signing more special contracts and thus confounded the Commission's bypass policy. The special contracts experience underscores ERAM's wide-ranging impact on utility risk and the difficulty of anticipating its full effect on new regulatory policies.

Encourages Financial Health

ERAM eliminates the potentially adverse effects of losses of sales from DSM and also automatically adjusts for other effects on sales, including weather fluctuations and the business cycle. This guarantee of revenues reduces the variability of earnings under unpredictable conditions and thus contributes significantly to utilities' financial health. The primary benefit of utility financial health is a lower cost of borrowing; whether such benefits accrue to ratepayers depends on the extent to which lower capital costs are reflected in rates.

While logically the stability of revenues guaranteed by ERAM must have improved the financial position of the California utilities, there are no empirical estimates of this effect. The CPUC clearly is proud that California's IOUs are generally financially healthy, and recent ratings by Wall Street advisors rank the CPUC quite favorably relative to other states.[9] Certainly, the aggressiveness of prudence reviews in the state and the interventionist regulatory approach temper the generosity of California's ratemaking mechanisms. ERAM would likely be a bigger boon to utilities in states without this tradition. The strength of utility opposition to eliminating California's ERAM confirms that ERAM serves their interests.

Redistributes Risk

It should be clear that ERAM alters the risk environment for an electric utility. If there is a fuel adjustment mechanism as in California, the company is already free of the risk of fuel cost escalation. If ERAM is also in place, the utility additionally receives blanket protection from the potential effect of sales variations. These two mechanisms together represent a powerful boon to an industry that cannot adjust its prices in the short run to reflect changing costs.

Consider ERAM's importance to a utility. One of the utility's major sources of risk, that projected sales to which revenues are tied will not materialize, is completely eliminated. Traditional issues such as weather fluctuations, recessions, and conservation impacts evaporate. ERAM and ECAC shift these risks to ratepayers because fluctuations automatically result in rate changes. There may be some benefit for the customer because a healthier utility can raise capital at a lower interest rate. The question of whether the resulting risk redistribution has benefitted ratepayers has not been answered in California.

Beyond the boundaries of regulated utilities, the issues are fuzzier. If ERAM deters competition in the industry, then the customer is denied both the benefits and the costs of competition. Many analysts in the industry believe that the introduction of competition will bring net positive benefits for customers (Primeaux 1986, Hamrin 1990, Mead and Denning 1990). It would be unfortunate if ERAM prevented this.

---

[9] Of all state public utilities commissions, California ranked in the top five in an investor rating system published by Merrill Lynch & Co. (Merrill Lynch 1991).

Other Effects

Removes Forecast Gaming Incentive. When ratemaking relies on forecasting, utilities can gain by effective sales forecast gaming. For the utility, an unforecast sale is as good as an additional sale, so it will attempt to underforecast sales before a rate case and promote sales after it. By guaranteeing that the utility will exactly recover its revenue requirement, the incentive to game with the sales forecast is eliminated. Further, ERAM makes this aspect of the ratemaking process less contentious and, potentially, more efficient. Forecasting is central to California regulation because the state uses a forecast test year and a forecast revenue requirement.

Creates Its Own Constituency. Like any other policy, ERAM has developed a life of its own in California. ERAM, a familiar and trusted friend that has provided significant benefits to the utilities, generated strong resistance to its proposed removal. Conservation groups, CPUC staff,[10] and other California state agencies have testified in favor of keeping ERAM. Also, as mentioned above, those groups, notably NUGs and large customers, who may have provided more opposition to ERAM have been mostly silent in California. Additionally, the argument that partial elimination of ERAM would be administratively burdensome has been powerful.

Encourages Innovative Ratemaking. One potential source of revenue variability that merits special attention is the consequences of imperfect, or experimental, ratemaking. If the base rate set in the GRC is incorrect due to forecasting error, the subsequent miscollection of revenues will accrue in the ERAM balancing account, so the utility is not affected by forecasting inaccuracy. It has been argued, especially by the utilities, that this ERAM benefit has improved the efficiency of the regulatory process by minimizing ratemaking disputes. With regard to experimental rates, the CPUC has been a national leader in introducing marginal-cost-based ratemaking, as well as interruptible tariffs and time-of-use rates. Such rates often rely on demand parameters that are difficult to forecast, such as on-peak demand. These ratemaking innovations might not have developed if the California utilities had not had ERAM to insulate them from forecasting error.

Requires Attrition. The argument that ERAM would be unfair without attrition is apparently valid. If a utility's base revenue were fixed, as it would be under ERAM *without* attrition, increases in costs

---

[10] CPUC staff are currently divided on the issue of whether to retain ERAM.

between general rate cases due to addition of customers or inflation would not be recouped. Adding of the attrition mechanism to ERAM reestablishes the link between costs and revenue.

Encourages Regulatory and Administrative Efficiency. With regard to eliminating the incentive to game forecasts, and eliminating fear of inaccurate ratemaking, it merits repeating that ERAM reduces the contentiousness of regulatory proceedings. Additionally, the ERAM account has been used as a convenient catchall account for minor ratemaking adjustments. Small corrections to rates are made by credits or debits to the ERAM account since these sums will eventually get rolled back into rates in future ERAM adjustments. ERAM, then, is useful for achieving regulatory efficiency.

Further, within the California regulatory context, ERAM is inexpensive and easy to administer. It does involve some additional accounting and creates some confusion in ratemaking due to different implementations by the utilities. However, from the conservationist perspective, these are trivial costs if the alternative is policing utility conservation programs in the field to ensure adequate performance.

Conclusions

Lessons for California

California's ratemaking and regulatory structure consists of unique procedures, policies, and mechanisms. Lessons from operating ERAM in the state can only be understood in this specific context.

ERAM protects the utilities from rate-of-return reductions that result from sales falling below forecasts. In the absence of ERAM, the erosion of earnings could be significant and would deter conservation programs. However, sales losses come from many sources, and it is not clear that current California policy goals justify protecting utilities from sales loss or gain. Conservation programs are different because they pose only downside risk to utilities. Unlike weather, the business cycle, and other causes of sales fluctuation, if conservation works, it affects sales only in one direction. While trying to correct for this phenomenon may be a reasonable policy goal, ERAM does it by providing blanket protection against all sales risk. There can be no clear guidance regarding the appropriate split of risk between utility and customer. Allowing the utility to bear additional risk has some negative effects, notably higher costs of borrowing and, as a result, higher rates. Conversely, shielding the utility from risk diminishes its commitment to operating efficiency and gives it an unfair advantage compared to NUGs.

Looking forward, ERAM may be a source of policy conflicts. The aim of separating large customers from small ones and applying ERAM to revenues only from the latter category has not been realized in California. Not only does such a plan impose significant administrative burdens, it represents a radical departure from traditional ratemaking, leading to complex, contentious cost allocation and oversight problems. ERAM is a mechanism that belongs to the era of the highly protected utility. Given that the financial health of the industry has improved, and that the overall trend in regulation is to make utilities more competitive, ERAM runs counter to wider regulatory goals of fostering competition, which could benefit ratepayers.

Finally, it should be emphasized that information about the importance of ERAM is limited. Even in the state of its birth after ten years' experience and study of its effects, knowledge of ERAM is limited to analysts who work directly on ratemaking matters. This lack of awareness has contributed to the adoption of poorly designed regulatory policies, such as the special contracts policy.

Lessons for Other States

Other states considering ERAM should recognize the need for some method to adjust the revenue requirement in response to load or customer growth. Otherwise, ERAM would simply be unfair to a growing utility. A revenue-indexing mechanism that allows partial recoupling of revenues to costs is one possible solution. It is essentially the strategy adopted by the CPUC in its attrition proceeding and by New York in its RDM mechanism. Annual rate cases are another possible way to partially recoupling revenues to costs. A simplified attrition mechanism based on the growth of customers in the utility's service territory is yet another possible solution, as described in Chapter 4. These mechanisms can be used by states that set rates using forecast test years and by states that use historical test years. States with historical-test-year ratemaking will need an attrition or indexing mechanism that adjusts the revenue requirement for the first year rates are in effect, because the revenue requirement will likely be an inaccurate measure of future costs. States that conduct future-test-year ratemaking should, by definition, have a reasonably accurate revenue requirement for the initial year, but adjustments to the revenue requirement will be needed for subsequent years.

Caution should be used when projecting results of California's ERAM experience to other states. No regulatory mechanism operates in a vacuum, and the importance of deviations in local conditions from those prevailing in California must be taken into account. The forecast test year, the central role of forecasting in ratemaking, the existence of ECAC, and the use of attrition, in particular, are conditions that have an impact on the operation of ERAM. California's proactive-yet-protective regulatory philosophy and its large regulatory staff are significant. The effects of an ERAM mechanism in another state might be quite different.

Acknowledgments

The work described in this study was partially funded by the Assistant Secretary for Conservation and Renewable Energy, Office of Utility Technologies, U.S. Department of Energy, under contract No. DE-AC03-76SF00098, and by the Universitywide Energy Research Group, University of California. The authors are also grateful for the assistance of the following: the editors of this volume; Joseph H. Eto, Charles A. Goldman, Edward P. Kahn, and Jon G. Koomey of the Lawrence Berkeley Laboratory; Prof. C. Bart McGuire of the School of Public Policy, U. C. Berkeley; David Moskovitz, Energy Regulatory Consultant; Ramesh Ramchandani, Pamela Thompson, Terry Mowrey, Carol Siegal, and James Weil of the California Public Utilities Commission; Ralph Cavanagh of the Natural Resources Defense Council; Terry Murray of Murray and Associates; Jim Cole of the California Institute for Energy Efficiency; Sam Swanson and John D. Stewart of the New York State Department of Public Service; Robert E. Burns of the National Regulatory Research Institute; and Nyla Marnay and Leslie Comnes of our respective homes. The opinions and views expressed in this paper are those of the authors and do not necessarily represent the views of the California Public Utilities Commission, or its Division of Ratepayer Advocates.

References

Ameer, Paul G. 1989. *Annual Energy Rate Study.* Fuels Branch, Division of Ratepayer Advocates, California Public Utilities Commission. August 25.

Barkovich, Barbara. 1989. *Regulatory Interventionism in the Utility Industry: Fairness, Efficiency, and the Pursuit of Energy Conservation.* Westport, Conn.: Greenwood Press.

California Electric Vehicle Task Force (CEVTF). 1989. *A California Plan for the Commercialization of Electric Vehicles.* July 11.

California Energy Commission and California Public Utilities Commission (CEC/CPUC). 1988. *Joint CEC/CPUC Hearings on Excess Electrical Generating Capacity*. P150-87-002. April.

California Public Utilities Commission. 1981. Decision 93887 in Application 60153. December 30.

California Public Utilities Commission. 1985. Decision 85-12-076. December.

California Public Utilities Commission. 1986. Order Instituting Rulemaking 86-10-001. October.

California Public Utilities Commission. 1987. Decision 87-05-071. May 29.

California Public Utilities Commission. 1988. Decision 88-12-041 in Rulemaking 86-10-001. December 9.

California Public Utilities Commission. 1989. Decision 89-05-067. May 26.

California Public Utilities Commission. 1990a. Decision 90-03-078 in Application 88-10-014. March 28.

California Public Utilities Commission. 1990b. Decision 90-12-128 in Application 89-04-001. December 27.

California Public Utilities Commission. 1991. Order Instituting Rulemaking 91-08-003 and Order Instituting Investigation 91-08-002. August.

Calwell, Chris J. and Ralph C. Cavanagh. 1989. *The Decline of Conservation at California Utilities: Causes, Costs and Remedies*. San Francisco: Natural Resources Defense Council.

Cavanagh, Ralph. 1988. "Responsible Power Marketing in an Increasingly Competitive Era." *Yale Journal on Regulation* 5. Summer.

Cole, James and Martin Cummings. 1990. "Making Conservation Profitable: An Assessment of Alternative Demand Side Management Incentives." In *Proceedings of the ACEEE 1990 Summer Study on Energy Efficiency in Buildings,* Vol. 5. Washington, D.C.: American Council for an Energy-Efficient Economy.

DeForest, Wayne R. and Paul L. Berkowitz. 1990. "DSM Competition: A New Regulatory Strategy." In *Proceedings of the ACEEE 1990 Summer Study on Energy Efficiency in Buildings,* Vol. 5. Washington, D.C.: American Council for an Energy-Efficient Economy.

Friedman, Lee S. 1991. "Electric Utility Pricing and Customer Response: The Recent Record in California." In *Regulatory Choices: A Perspective on Developments in Regulatory Policy,* R. J. Gilbert, ed. Berkeley: U. C. Press.

Gilbert, Richard J. 1991. "Issues in Public Utility Regulation." In *Regulatory Choices: A Perspective on Developments in Energy Policy,* R. J. Gilbert, ed. Berkeley: U. C. Press.

Hamrin, Jan. 1990. "Pricing a New Generation of Power: A Report on Bidding." In *Competition in Electricity: New Markets and New Structures*, J. Plummer and S. Troppmann, eds. Arlington, Va.: Public Utilities Reports.

Marnay, Chris. 1989. *Special Electricity Contracts in California*. Universitywide Energy Research Group Report #242. Berkeley: University of California.

Marnay, Chris and G. Alan Comnes. 1990. *Ratemaking for Conservation: The California ERAM Experience*. LBL-28019. Berkeley: Lawrence Berkeley Laboratory. March.

Mead, Walter and Mike Denning. 1990. "New Evidence on Benefits and Costs of Public Utility Rate Regulation." In *Competition in Electricity: New Markets and New Structures*, J. Plummer and S. Troppmann, eds. Arlington, Va.: Public Utilities Reports.

Merrill Lynch & Co. 1991. "Utility Industry Opinions on Regulation." New York: Merrill Lynch & Co. July 2.

Messenger, Michael. 1989. *Will Electric Utilities Effectively Compete in Markets Without a Profit Motive? An Analysis of the Last Decade of Energy Conservation Programs in California*. Sacramento: California Energy Commission.

Murray, Terry L. 1991. "Direct Testimony and Exhibits." NYPSC Case Nos. 29327 et al., 89-E-152, 89-E-153, and 89-G-15 (Niagara Mohawk Power Corporation).

National Association of Regulatory Utility Commissioners. 1988. *NARUC Bulletin*. August 8.

New York Public Service Commission (NYPSC). 1989. Opinion 89-29 in Case 89-E-176. September 12.

New York Public Service Commission (NYPSC). 1990. "Opinion and Order Determining Revenue Requirement and Approving Revenue Decoupling Mechanism." Opinion 89-24 in Cases 89-E-175 and 89-E-176. September 26.

New York Public Service Commission (NYPSC). 1991. Opinion 89-37(D) in Case 29327 *et al*.

Pearl, Lewis J., and Mark Luftig. 1991. "Financial Implications to Utilities of Third-Party Power Purchases." *The Electricity Journal* 3. November.

Phillips, Charles F., Jr. 1988. *The Regulation of Public Utilities: Theory and Practice*. Arlington, Va.: Public Utilities Reports.

Plummer, James, and Susan Troppmann, eds. 1990. *Competition in Electricity: New Markets and New Structures*. Arlington, Va.: Public Utilities Reports.

Primeaux, Walter. 1986. *Direct Electric Utility Competition: The Natural Monopoly Myth*. New York: Praeger.

Radford, Bruce W., ed. 1988. *Rate-Making Trends in the 1980s.* Arlington Va.: Public Utilities Reports.

Reid, Michael W. and Edward M. Weaver. 1991. *The Michigan Incentives Study for Electric Utilities: Phase I Final Report.* Washington, D.C.: Barakat & Chamberlin, Inc. April.

Sissine, Fred. 1989. *Making Conservation Profitable: Issues for Regulation and Evaluation.* Briefing paper prepared for the Plenary Panel of the Fourth Energy Program Evaluation Conference, Chicago. Washington, D.C.: Congressional Research Service, Library of Congress. August 24.

Statewide Collaborative Process (SCP), 1990. *An Energy Efficiency Blueprint for California.* San Francisco: California Public Utilities Commission. January.

Woo, Shirley A. and Roger J. Peters. 1989. Comments of Pacific Gas and Electric Company in Response to Ordering Paragraph 4 of Commission Decision 88-12-041. San Francisco: Pacific Gas and Electric Company. January 13.

Ziering, Mark A. 1986. "Risk, Return, and Ratemaking: A Review of the Commission's Regulatory Mechanisms." Policy and Planning Division, California Public Utilities Commission. October 1.

Chapter 4

Revenue-per-Customer Decoupling

David Moskovitz and Gary B. Swofford

Introduction

Chapter 3 described how California's Electric Revenue Adjustment Mechanism (ERAM) decouples utility profits from sales. ERAM is useful in states that rely on future-test-year (FTY) ratemaking. This chapter describes another decoupling method that can be used in either FTY or historic-test-year (HTY) jurisdictions.

Principles of Historic- and Future-Test-Year Ratemaking

Historic-test-year ratemaking assumes that a recent historic period, adjusted for normal conditions and updated for known and measurable changes, provides a reasonable basis on which to establish the relationship between prices and revenues compared to expenses and investment.

At the end of a rate case, regulators decide how much revenue the utility should have been allowed to collect from customers during the historic test year. Then regulators establish electricity prices that, if charged during the historic period, would have collected that revenue from customers. Note that regardless of any adjustments made during the course of the rate case, the allowed revenue is *not* the same as the revenue the company will collect in subsequent years when the new rates will be in effect.

When prices have been set, utility revenues are solely a function of sales. Because sales levels in the rate year (the first year the new rates are in effect) and beyond will usually be higher than during the historic test year, the actual revenue collected will normally be higher than allowed revenue; the amount depends on the exact sales level.

In theory, revenue from increased sales will not result in higher profits because higher revenues will be offset by cost increases. Historic ratemaking assumes a proportional relationship between sales growth and expenses. This critical assumption, rarely tested or questioned, is a key justification for "coupling" profits to sales; however, a close examination of this cost/sales assumption casts doubt on its accuracy.

While the mechanics of FTY and HTY ratesetting differ, the fundamentals are the same. Historic-test-year ratesetting depends on a historic period to establish the revenue/cost relationship and prices. Once approved by regulators, prices stay unchanged for the indefinite future. Future-test-year ratesetting relies on a forecasted test period to do the same thing: establish a revenue/cost relationship to establish prices that are allowed to operate into the indefinite future.

In addition, unlike HTY procedures, the FTY process includes an estimate of the revenue the utility will receive the first year rates are expected to be in force. This forecast of revenue makes decoupling with an ERAM-type mechanism possible (see Chapter 3). The lack of a revenue forecast in a historic-test-year jurisdiction precludes using ERAM-type approaches. The revenue-per-customer approach described in this chapter provides a means to decouple profits from sales in both HTY and FTY jurisdictions, which is important because most states use a HTY ratesetting process.

Development of the RPC Approach

The revenue-per-customer (RPC) approach is an alternative decoupling option, developed especially for states using HTY ratesetting. In these jurisdictions, severing the sales/revenue link means that a new mechanism is required to allow HTY revenues to grow and keep pace with cost growth and increased energy service demands in the rate year and beyond. If revenue growth is "recoupled" to something other than sales levels, it must be to something that makes economic and policy sense.

It was first suggested in *Profits and Progress Through Least- Cost Planning* (Moskovitz 1989) that permitting revenues to grow in proportion to customer growth instead of sales would effectively decouple profits from sales.

Linking revenue levels to customer growth is equivalent to allowing the utility to recover a fixed amount of revenue per customer. If a utility is allowed a specific amount of money for meeting a consumer's energy service needs, the utility will have an incentive to meet those needs at the lowest possible cost. The difference between the revenue the utility receives from the customer and the cost incurred by the utility to serve the customer is the utility's profit. The utility's financial incentive is clearly consistent with least-cost planning.

In practice, it would be necessary to determine if the customer/ cost relationship using the RPC approach produced a "reasonable" result compared to traditional sales-based regulation. This outcome would permit the RPC approach to satisfy the regulatory test of reason ("just and reasonable") and protect consumer interests.

In 1990, Puget Sound Power and Light Company (Puget Power), the largest investor-owned utility in the state of Washington, was encouraged by the Washington Utilities and Transportation Commission (WUTC) to explore the feasibility of the RPC decoupling approach.

The first step in developing an RPC mechanism was to divide Puget Power's costs into two broad categories, defined in Washington State as "base" and "resource" costs. Other jurisdictions often label these costs "non-fuel" and "fuel" costs. This step was necessary in Washington because at the time, no fuel or purchased power clause existed.[1]

Resource costs include fuel, purchased power, and conservation-related costs. As is the case in most other states, Puget Power will now recover these costs separately through an annual adjustment mechanism that operates independently from the RPC mechanism. All remaining costs, which most states call "non-fuel" costs, were classified as base costs.

An equivalent base/resource or non-fuel/fuel cost division occurs under ERAM. Fuel and similar costs are treated separately using a fuel clause. Only non-fuel, or base costs, are subject to the ERAM process.

Once the relevant costs were identified, Puget Power analyzed the historic relationship between its base costs and kilowatt-hour sales, and compared these costs with the number of customers.

Data on sales (both weather and non-weather adjusted), customer count, and base costs were collected for the most recent 15-year period. As a first step, simple regression analyses identified the long-term historical relationships between base costs (the dependent variable) and sales and customer count (the independent variables). More sophisticated regressions designed to test and adjust for autocorrelation

---

[1] Although lack of a fuel or purchased power clause was an issue, the Washington experience showed that it was not a barrier to adapting the RPC approach.

Table 4-1. Regression Analysis Results for 1972 Through 1989, Simple and Time Adjusted	
Regression	$\mathbf{R^2}$
Sales vs. base costs (simple)	.91
Customers vs. base costs (simple)	.96
Sales vs. base costs (time adjusted)	.97
Customers vs. base costs (time adjusted)	.99

were also performed.[2] The results of the latter analyses are called "time adjusted."

The results of these analysis are shown in Table 4-1.[3] Both on a simple and time-adjusted basis there is a reasonably strong statistical correlation between sales and base cost growth. There is, however, an even more substantial relationship between customer levels and movement in actual base costs.

Additional studies assessed the statistical relationships between annual changes in base costs, and annual changes in both sales and customer levels. Analyses of yearly variations are useful because one- and two-year time intervals typify the periods between general rate cases. These studies showed that over a short period of time, utility base costs were not directly related to sales growth or customer growth.

The lack of a short-term relationship between base costs and sales levels is consistent with experience in rate cases. For example, in FTY jurisdictions (except California) utilities usually argue that sales will be low, while consumer advocates argue that sales will high. Both parties know that the utility's base revenue requirement will not be affected by which sales forecast is adopted. With the numerator of the price-setting formula fixed (price = revenue/sales), prices will be highest with the lowest sales estimate. The utility, of course, prefers high prices because without decoupling, actual revenues depend on actual, as opposed to predicted, sales.

---

[2] Regression analysis performed on time series data frequently exhibits autocorrelation. This means that the apparent relationship between the dependent variable (costs) and the independent variable (customers or sales) may be distorted by the effect of time. Standard statistical techniques can measure and correct for this imprecision.

[3] The crucial value in this table is ("adjusted") R^2, which represents the proportion of the variability in base cost "explained" by the impact of the independent variable, customers or sales. The closer R^2 is to 1.0, the stronger the statistical relationship between the dependent and independent variables.

Table 4-2. Central Maine Power Company Regression Analysis results	
Regression	**R^2**
1972–1989	
Sales vs. costs	.98
Customers vs. costs	.98
Annual changes	
Sales vs. costs	.12
Customers vs. costs	.22
Biannual changes	
Sales vs. costs	.33
Customers vs. costs	.42

The analysis of historical cost, sales, and customer count data leads to two conclusions:

- In the long run the relationship between cost and customer growth is stronger or no worse than the corresponding relationship between costs and sales.

- The short-run analysis of year-to-year changes in sales vs. base costs shows no statistically significant relationship. Yet, as previously described, the assumed existence of a strong correlation between these two factors is the foundation of traditional sales-based regulation.

Similar statistical studies have now been performed on data for other utilities. Relationships between utility costs, sales growth and customer base have been analyzed for Potomac Electric Power, Central Maine Power, Public Service of Colorado, and the New England Electric System. Although the results differ among the utilities, the general conclusions are the same. Each study has shown that over long periods of time, "decoupling" revenue from sales and "recoupling" revenue to customer growth has a statistical basis that is *at least as good as the existing system*. For example, the short-run analysis of Central Maine Power showed a significantly stronger relationship between customer and cost growth than between sales and cost growth. The R^2 results from 1972 to 1989 were similar to those of the Puget Power study. The results for Central Maine Power are shown in Table 4-2.

Building on this statistical analysis of historical costs, Puget Power next analyzed the effect of an RPC decoupling plan using a financial simulation model to forecast revenues and costs under conventional regulation and under a hypothetical RPC approach. The analysis established that the RPC approach would not produce a windfall for consumers or Puget Power.

Mechanics of RPC

General rate cases proceed according to current methods. No additional conflicting issues are injected into the process and the timing of rate cases continues on an "as needed" basis. The only added requirement is the establishment of the utility's average allowed base (nonfuel) revenue per customer. The data required to construct this measure are already available in the ordinary course of rate cases in the form of allowed revenues divided by the average-test-year customer count. Revenue per customer is computed by dividing allowed revenues by the test-year number of customers:

$$\frac{\text{Test-Year Allowed Revenues}}{\text{Test-Year Customers}} = \text{Revenue per Customer}$$

The allowed revenue per customer remains fixed until the next general rate case. The need for periodic general rate cases will continue. The RPC decoupling method is not designed to change the length of time between general rate cases. The utility remains free to initiate a general rate case if its financial condition requires it. Likewise, the regulatory commissions, public advocate, or others may initiate a rate proceeding if utility earnings are perceived to be excessive.

On April 1, 1991 Puget Power's proposal was approved by the WUTC. A similar approach was adopted in Maine on April 12, 1991.

Under the RPC plan, rate design issues are unaffected. Customer, energy, and demand charges are determined according to previous practice. During the rate year and thereafter the utility bills for service and collects revenues. At the end of each year, utility revenues are compared to the product of the allowed revenue per customer and the average number of customers during the year. Any disparity between allowed and actual revenues is reflected as a surcharge or refund to customers during the following year.

Fine Tuning

The RPC approach is flexible enough to be adapted to a variety of circumstances. In general, RPC can be used if the customer base is increasing or decreasing. The relevant consideration is whether changes in customer numbers differ significantly from changes in sales levels. In other words, are kWh sales per customer relatively constant despite customer numbers? Utilities with a declining customer base generally experience declining sales. The issue for these utilities is whether their revenues under RPC would differ significantly from the revenues they would receive under conventional regulation. The RPC approach has enough flexibility to address this issue.

In Washington and Maine, revenue growth is based on the following equation:

$$\text{Base Revenue Growth} = K * \text{Customer Growth}$$

Where K is fixed and equals 1.0.

Under traditional regulation, revenue growth is defined by a similar equation:

$$\text{Base Revenue Growth} = K * \text{Sales Growth}$$

With traditional regulation "K" is roughly equal to 1.0, but it is not fixed or prespecified. The "K" factor is instead an implicit by-product of rate design decisions and the actual pattern of sales growth. For example, replacing a declining block rate with an inverted block rate will alter the relationship between sales and revenues and implicitly change the "K" factor.

Although the "K" factor established for both Puget Power and Central Maine Power was 1.0, other circumstances may call for a different "K" factor above or below 1.0. For example, a utility experiencing sales growth and revenue growth that substantially exceeds customer growth would likely oppose RPC unless the plan produced revenues approximating the revenues it would expect to receive under conventional regulation. A "K" factor greater than 1.0 could be used to address this issue.

RPC Compared to ERAM

As discussed in Chapter 3, ERAM decouples profits from sales by establishing a fixed amount of base revenue that the utility will be allowed to recover. However, the forecasted test year establishes the allowed revenue for only the first twelve months. In the future, either the utility will go though another general rate case to establish a new forecasted test year, or some abbreviated method will be used to adjust the most recent test year to reflect changes in conditions. California and New York State have adopted ERAM-type mechanisms with short-cut "attrition" proceedings to adjust (generally increase) the allowed revenue each year until the next general rate case.

RPC can also be used in an FTY jurisdiction. The forecasted base revenue requirement is still be determined in the rate case, but the utility's actual revenue is not fixed or guaranteed, as is the case with ERAM. Instead, allowed revenue per customer is computed in the rate case by dividing forecasted revenue requirement by the projected number of customers. Allowed revenue in the rate year is determined by multiplying the allowed revenue per customer by the actual number of customers. Thus, under ERAM, allowed base revenue is determined in a rate case or between full rate cases in annual "attrition" proceedings. Under RPC, allowed revenue is determined by the change in the number of customers. The result is decoupling that is as effective as ERAM without the need for annual rate cases or a separate attrition mechanism.

Most states rely on HTY ratesetting, and consumer advocates and most regulatory commissions strongly resist the adoption of FTY practices. Future-test-year ratesetting depends on forecasts of sales and the costs of each aspect of the utilities' business. Principal barriers to wider adoption of FTY ratesetting are the level of resources needed for detailed forecasting and budgeting, and concern that utilities control the data and can dominate the ratesetting process.

Under RPC, the recovery of base revenue is independent of sales levels, but revenues are neither forecast nor fixed. Recovery of non-fuel or base costs depends on a customer count, a variable that is much less under the company's control and apparently more closely related to base costs than sales.

By decoupling profits from sales, the RPC approach removes the disincentives for utility DSM activities as well as other more subtle disincentives. For example, under the traditional system there are disincentives to utility adoption of many innovative rate designs. Utilities justifiably expect that time-of-use, seasonal, and inverted block rates and hook-up fees would all cause significant earnings and revenue losses as customers respond to a more accurate price structure. Decoupling removes this disincentive, because earnings and revenues are independent of customer response to new prices.[4]

Does RPC Provide Positive Incentives?

Like other decoupling approaches, the revenue-per-customer approach does not in itself produce positive incentives for DSM or rate design changes. Nevertheless, RPC, like ERAM, retains the existing cost-

---

[4] Note that DSM-specific lost revenue adjustments do not offer this benefit.

cutting incentives produced by regulatory lag. Revenues are tied to customer growth, but otherwise no adjustments to rates or allowed revenue per customer are made to reflect changes in costs. Because profits are the difference between revenues and costs, the utility's ability to economize in any area of non-fuel costs directly benefits shareholders.

Potential Problems

A number of questions have been raised during the course of implementing the RPC decoupling plans in Washington and Maine.

Customer Size and Mix

The RPC decoupling plans implemented in Maine and Washington are based on aggregate customer counts that include all classes of customers. In both states questions were raised concerning the implications of counting residential, commercial and industrial customers equally.

A mitigating factor is that the RPC computation refers solely to *base* costs. The *resource* cost recovery continues to be based on sales levels, so the *total* allowed revenue from the average industrial customer will be much greater than that for the average residential customer.

In addition, RPC does not affect customer billing. It is simply a ratemaking mechanism to determine a utility's allowed base revenue. In contrast, in a state with an FTY and ERAM, regulators set allowed base revenue using cost projections. Any significant change from projections, such as the unexpected addition of a large customer to the system, is assumed to change revenues, not costs. With ERAM, the incremental base revenue is returned to customers. Similarly, with RPC the addition of a large new customer would have a slight increase in allowed base revenues. Any excess base revenue is returned to customers.

The RPC decoupling approach could be implemented using class-specific revenue and customer data. Mathematically, however, the aggregated and customer class-specific approaches produce the same allowed revenue if the utility's mix of residential, commercial, and industrial customers remains constant. Given the size of Puget Power and Central Maine Power, and the relatively short time between general rate cases, significant shifts in customer mix were not expected.

Regulators in both Washington and Maine considered using separate revenue-per-customer calculations for residential, commercial, and industrial customers. Constructing a workable mechanism was unjustified for either utility.

Further efforts in these and other states may lead to adoption of a customer-class-based RPC mechanism. Utilities that have few large industrial customers may prefer this approach. It should be noted that under conventional regulation the addition or loss of a very large customer to the service territory of a small utility often creates the need for a rate case.

New Customers and Gaming

With revenues tied to the number of customers, the utility will obviously have an incentive to attract new customers. An additional, and possibly perverse, incentive may also be created to increase the apparent number of customers by counting them in a way that would be more advantageous to the utility.

With respect to the incentive to attract new customers, the same incentive exists under conventional regulation except that under the current system, the utility receives more revenue from larger and less energy-efficient customers. With RPC the utility has an incentive to attract smaller and more energy-efficient customers who can be inexpensively served. Serving these new customers will be more profitable because the incremental revenue is unrelated to additional energy generation or consumption.

To guard against gaming opportunities such as sub-metering master-metered locations, or adding meters to street lights, fax machines, and phone booths, both Washington and Maine adopted detailed written definitions and procedures for counting and verifying customers. These are important and necessary elements of any RPC decoupling plan.

Risk Allocation

Decoupling, whether through ERAM or RPC, also shifts the financial exposure due to weather-related revenue volatility from utility shareholders to consumers. Under RPC decoupling, unusual weather conditions will produce variations in kWh sales but will not affect the number of customers. If the number of customers is unaffected, the utility's revenues will also be unaffected.

Under RPC, the impact of sales volatility caused by economic conditions is more difficult to project. Regional economic conditions will undoubtedly affect customer growth, although likely to a lesser extent than the economy affects kWh sales. If this is case, RPC decoupling shifts part of the business cycle risk from utility shareholders to consumers.

Decoupling may increase the level of risk perceived by utility shareholders because it represents a relatively untested change in regulatory practice.

The net change in the utility's risk will ultimately be reflected in its cost of capital. Regulators considering the adoption of decoupling in general, or RPC in particular, should consider if decoupling will affect the utility's cost of capital. Regulators should also judge whether any adjustment should be made coincident with adopting decoupling or deferred until the next rate case, when changes in investor perceptions may be easier to measure.

In both Washington and Maine there were arguments about the net change in shareholder risk and whether the allowed rate of return should be adjusted concurrent with decoupling. Neither state commission made such an adjustment.

The Impact of RPC on Puget Power

The decisive question is whether regulatory innovations such as RPC will produce tangible impacts on specific utilities and their pursuit of least-cost planning. While a full assessment of the effects of regulatory reforms in Washington may be premature, there is concrete evidence of positive change.

Figure 4-1 shows actual DSM spending by Puget Power from 1983 to 1991 and budgeted DSM spending for 1992. Before regulatory reform, revenues were based on sales, and DSM programs had the effect of lowering sales and constraining profits. As a result, while Puget Power may have had other justifications for investing in conservation, its spending was limited by a rational response to existing disincentives.

Puget Power's annual investment in DSM was relatively constant for eight years. Investment in DSM since the initial development of the decoupling plan has doubled, and DSM savings have tripled (Figure 4-2). Projected spending and DSM savings represent more than 6% of revenues and about 50% of projected load growth. By either measure, Puget's DSM programs are now among the most aggressive in the nation.

Investor response to the RPC mechanism appears to be favorable. Between January and the April 1, 1991 ruling by the WUTC approving the decoupling plan, the value of a common share of Puget's stock increased by about 8%, compared to an increase of about 4% for Standard & Poor's Electric Utility Index.

From the date of the WUTC ruling to November 1, 1991, Puget's common stock rose by 15% compared to an increase in the S&P Index

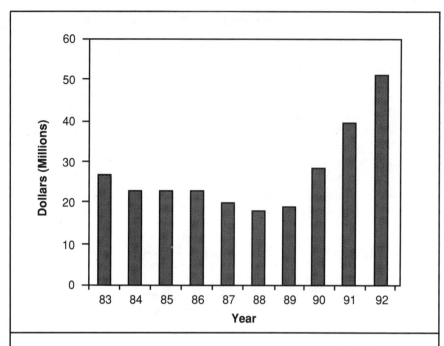

Figure 4-1. Puget Power's DSM Investment in Millions of Dollars (1992 data projected)

of about 7%. Figure 4-3 shows the performance of Puget's common stock relative to the S&P Index.

Securities analysts apparently perceive a benefit from the new mechanism; the Value Line Investment Survey concluded that "Puget Power has received a good grade order from the Washington-Regulators" (Value Line 1991a).[5] Whether this positive reaction persists will depend on the long-term implications of the overall regulatory reforms. The short-term positive reaction of the financial community, however, will reinforce Puget Power's DSM efforts and suggests that investors do not believe RPC will be detrimental to financial performance.

_____

[5] In the case of Central Maine Power, Value Line concluded that "Central Maine will benefit from a new revenue mechanism approved by Maine regulators" (Value Line 1991b).

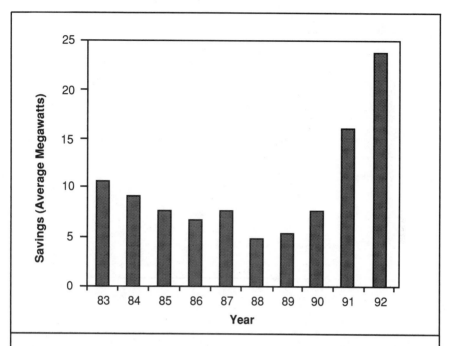

Figure 4-2. Puget Power's DSM Savings in Average Megawatts (1992 data projected)

Conclusion

Assigning a definite cause-and-effect relationship to specific regulatory measures and subsequent utility performance is difficult. Utilities operate in a complex and shifting economic environment in which the immediate effects of regulatory intervention are often ambiguous. Nevertheless, we can conclude that Puget's recently expanded DSM activities are the product of the new RPC decoupling plan and improved DSM cost recovery.

RPC, however, is only one element of a comprehensive regulatory reform plan under consideration in Washington. The RPC approach effectively decouples profits from sales, and the resource portion of the plan assures recovery of DSM costs. Although the RPC plan does not provide positive DSM incentives, an incentive plan applying to 1991 DSM performance is pending Commission approval.

The revenue-per-customer approach seems to be a promising tool

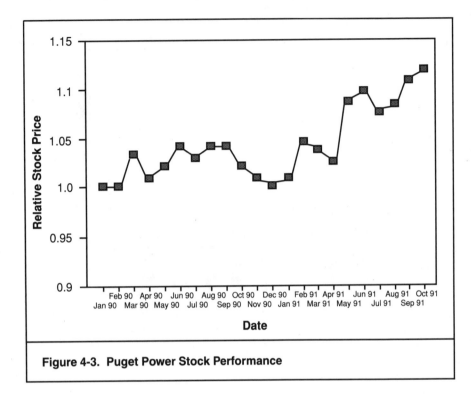

Figure 4-3. Puget Power Stock Performance

with which to pursue the regulatory objective of stimulating least-cost planning by decoupling utility revenues from sales. RPC has the advantage of being relatively simple. It is adaptable to a wide range of utility circumstances. The response by management and the financial community to its initial application has been encouraging. Finally, unlike ERAM, RPC decoupling can be used in a historic- or future-test-year jurisdiction.

References

Moskovitz, David. 1989. *Profits and Progress Through Least-Cost Planning*. Washington, D.C.: National Association of Regulatory Utility Commissioners.

Maine Public Utilities Commission. 1991. Investigation of Incentive Rate Proposal of Central Maine Power Company—Joint Report Regarding Interim Incentive Rate Plan. Docket No. 90-085.

Value Line Investment Survey. 1991a. New York: Value Line Publishing Company. May 31.

————. 1991b. September 20.

Washington Utilities and Transportation Commission. 1991. Third Supplemental Order in the Matter of the Petition of Puget Sound Power & Light Company for an Order Approving a Periodic Rate Adjustment Mechanism and Related Accounting. Docket No. UE-901194-P.

Chapter 5

Ratebasing of DSM Expenditures

Michael W. Reid

Introduction

Ratebasing is often mentioned in two contexts when regulatory incentives for DSM are discussed: as a method to recover DSM program costs, or as an approach to stimulating greater DSM investment by utilities. This chapter explores the concept of ratebasing and analyzes its usefulness for program cost recovery and as a stimulus to DSM.

What Is Ratebasing?

Ratebasing is an accounting and cost-recovery method applied to certain categories of utility expenditures. In general, a utility expenditure that is chargeable to customers is either *expensed* or *ratebased*. An outlay that is expensed is treated as an operating cost in the year it is incurred, and, provided its recovery has been authorized by the regulatory commission, is recovered concurrently from ratepayers.

When an outlay is ratebased, it is included in the utility's net investment, or *rate base*, that is used in the ratemaking process. This treatment has two effects:

1. The utility recovers the expenditure over two or more years (the amortization period), with equal amounts collected through rates each year.

2. The utility earns a return on the unrecovered (unamortized) portion of the expenditure during each year of the amortization period. The return, computed as a percentage of the unamortized balance, is also collected through rates.

Ratebasing is the cost-recovery method normally applied to utilities' investments in generation, transmission, and distribution equipment. It was conceived partly to apportion to customers the costs of investments required to provide energy services. A generating plant, for example, might be ratebased for 30 years, creating a cost stream that is passed on to ratepayers while the plant is expected to be in service.

Ratebasing also serves to compensate the utility for the use of capital provided by investors and creditors. The utility's total rate base is the estimated dollar amount it has invested that enables it to meet its obligation to serve. Most of this investment is in property and equipment that provides energy services, but other components, such as working capital, are usually included. The allowed components of rate base vary from state to state.

The allowed rate of return on the rate base is customarily set at the regulators' estimate of the utility's cost of capital, considering equity and debt sources. In principle, allowing a utility to earn a return on the rate base equal to its cost of capital provides shareholders the opportunity to earn a fair profit and enables the utility to attract capital.

Utilities' expenditures on demand-side management (DSM) are *not* routinely included in the rate base, even when programs produce long-lived streams of benefits. Instead, DSM expenditures are expensed; that is, they are recovered from ratepayers in the same year they are incurred, without a return.

If a DSM expenditure is ratebased, however, it is treated like generation, transmission, and distribution assets: recovery is spread over several years, during which time the utility earns a return. In some states, DSM expenditures might literally be included in the rate base. In other states, DSM can effectively be given rate base treatment without altering the book value of the rate base. For analytical purposes, any treatment that amortizes the expenditure and allows the utility to earn a return during the amortization period may be considered ratebasing.

Note that ratebasing is distinct from another regulatory concept, "base rates," which refers to the rates paid by utility customers before fuel adjustments and other surcharges are added. This unfortunate similarly in terminology often confuses discussions of ratebasing. Expensed costs as well as ratebased costs can be recovered in base rates. Both types of costs can also be recovered in surcharges.

Reasons for Ratebasing DSM

Ratebasing of DSM has been suggested by various energy conservation and consumer advocates, utilities, and regulators. Their arguments in support of ratebasing can be grouped in four categories.

"Leveling the Playing Field"

A central tenet of integrated resource planning (IRP) is that the supply and demand sides should be treated equivalently—or, as is often stated, they should compete on a "level playing field." Applying different accounting and ratemaking treatments to resource options appears to thwart this principle: if DSM and supply-side investments have different financial effects, how can utilities be expected to consider them equivalently when selecting resources? Ratebasing DSM, therefore, is seen as a means to further the goals of IRP and achieve the energy and economic efficiency gains IRP promises.

Offering a Financial Incentive

When DSM expenditures are expensed, they are simply passed on to ratepayers and do not contribute to utility earnings. Ratebasing, on the other hand, increases the base used to compute the utility's authorized earnings. Ratebasing is often considered a financial incentive for DSM, since it creates the opportunity to add to utility earnings: "Allowing a return on conservation investments would put conservation on an equal footing with supply options and provide a financial incentive for utilities to invest in conservation programs" (Chase 1987).

A variation on this argument is that ratebasing of DSM is needed to offset declines in the supply-side portion of the rate base. As existing plant and equipment are depreciated, a utility's total rate base will decline unless the reductions are matched by new investments. Some utility executives believe that a declining rate base will be viewed unfavorably by investors, harming market value and increasing the cost of capital. The options for maintaining or expanding the rate base are to build new plants—often an unrealistic option for economic, environmental, or political reasons—and/or to put the expanding DSM budget into the rate base. "If we're not going to be allowed to build a powerplant, we need some other opportunities to grow," said one utility executive in support of DSM ratebasing (Krzos 1988).

Providing Efficient Price Signals

An objective of utility regulation is to provide consumers with efficient price signals—prices that accurately reflect the costs of energy consumption. This means that the cost of a long-lived resource should be apportioned over its life rather than borne by consumers in a lump sum. Treating the cost as a lump sum, as expensing does, makes the resource appear more costly than if the charges were spread over several years and could lead to underconsumption.

A similar argument is based on the principle of intergenerational equity. If a DSM program is expensed, its full costs are borne by today's ratepayers. If the program provides several years' worth of benefits, future generations of customers receive those benefits without bearing any of the costs. Equity is served when all the program's beneficiaries share the costs.

Promoting Rate Stability

Rapid expansion in a utility's DSM budget could lead to "rate shock," a sudden rate rise, if the costs are expensed. Regulators aim for continuity in rates to avoid public backlash and to reduce the possibility that large customers will go off-system, which could cause rate increases for remaining customers. Because it spreads the costs over time, ratebasing is considered a means to minimize rate shock from expanding DSM programs.

Use of Ratebasing for DSM

While not widespread, ratebasing of DSM has been authorized in several states. This section describes instances of DSM ratebasing and illustrates significant variations in approach. Three important dimensions of ratebasing policies are discussed: the eligible expenditures, the rate of return allowed, and the amortization period.

Types of DSM Expenditures Eligible for Ratebasing

Utilities' DSM expenditures often include purchases of load-control equipment, such as radio-controlled switches installed on customers' air conditioners and water heaters. Such equipment remains the property of the utility and may be considered an extension of the distribution system. DSM expenditures of this type are often ratebased.

Utilities have also been allowed to ratebase financial assets associated with DSM programs. In 1978 Pacific Power & Light (PP&L) was authorized by the Oregon commission to ratebase loans to residential customers for weatherizing their homes. The loan balances were allowed to remain in the rate base until repaid by the customer, with the full balance due on the sale of the home (Krasniewski and Murdock 1980).

Following Oregon's lead, the Idaho commission authorized ratebasing similar loan programs in 1979 and 1980. But Idaho went further in allowing PP&L to also ratebase the cost of water heater wraps it gave customers. This decision broke new ground because the expenditure for the wraps was not clearly an investment (like the loans), and the utility did not retain ownership of the equipment. The fact that the water heater wraps were expected to save energy for several years was sufficient for the commission to authorize ratebasing (Idaho Public Utilities Commission 1979 and 1980).

Policies allowing utilities to ratebase DSM equipment over which they do not retain ownership have been adopted in other states. For example, a 1988 Maine statute states that

> 'Electric Plant' includes, but is not limited to, fixtures and personal property paid for in whole or in part by the utility on the premises of any of its customers as long as this property has been found by the Commission to constitute a cost effective investment in conservation or load management. . . . The Commission may in its discretion include in the utility's rate base, and permit a fair return thereon, any electric plant to the extent paid for by the utility which constitutes a cost effective investment in conservation and load management and which was installed on the premises of a customer. . . . (Maine 1987)

Similarly, a 1988 order from the Massachusetts commission declared that

> Electric companies can earn a return on C&LM [conservation and load management] equipment and materials, along with related capitalizable labor and administrative costs, where such expenditures will provide long-run benefits to ratepayers. . . . When an electric company does not own these capital assets, it is appropriate to allow the company to amortize the expenditures over the life of the asset and to earn a return on the unamortized balance (Massachusetts 1988).

Substantial portions of many utilities' DSM budgets are allocated to financial incentives, especially rebates paid to customers for purchases of energy-efficient equipment. In some states incentives can be rate-based. Wisconsin initiated this practice in 1986 when it directed Wisconsin Electric Power Company (WEPCO) to undertake a massive $84 million DSM program. Seventy-three million dollars of the total, intended to be used for rebates, interest rate buy-downs, loan guarantees, and direct loans, was ratebased; the balance, predominantly for program administration and research, was expensed (Wisconsin Public Service Commission 1986).

The most expansive view of ratebasing is that it should be allowed for any expenditure made to create or support a long-lived DSM resource. A few utilities have successfully petitioned for such treatment. In Washington State, for example, Puget Sound Power & Light has been allowed to ratebase most of its DSM budget, including conservation-related advertising, informational, and educational expenditures. In 1990 Southern California Edison received authorization to ratebase a major portion of its DSM budget; programs given ratebase treatment were those expected to provide quantifiable long-term resource benefits (California Public Utilities Commission 1990).

These examples collectively show that ratebasing of DSM expenditures can be narrow or broad in scope. When ratebasing is allowed, both legislation and regulatory decisions govern whether a portion of DSM expenditures or the entire DSM budget will be eligible.

Allowed Rate of Return on Ratebased DSM

Two approaches have been used to set the allowed rate of return (ROR) on ratebased DSM:

- Applying the same ROR used for other components of the utility's rate base. This value is usually the regulatory commission's estimate of the utility's weighted-average cost of capital.

- Allowing the utility to earn a bonus ROR, a return greater than that authorized on other items in the rate base. Bonus returns are generally expressed as a percentage adder or "kicker" to the equity portion of the utility's ROR.

The former approach is consistent with the principle of treating all resource options equitably, since both supply- and demand-side investments earn the same return during their amortization periods. Some of the early instances of ratebasing, such as PP&L's weatherization programs in Oregon and Idaho, were treated this way. SCE's 1990 ratebasing plan is a more recent example.

Greater-than-normal returns have been allowed in several states in an effort to provide an incentive for DSM, and in some cases other resources, such as cogeneration and renewables. A 1980 Washington State statute, for example, instructed the commission to

> . . . adopt policies to encourage meeting or reducing energy demand through cogeneration . . . , measures which improve the efficiency of energy end use, and new projects which produce or generate energy from renewable resources. . . . These policies shall include but are not limited to allowing a return on investment . . . which return is established by adding an increment of two percent to the rate of return on common equity permitted on the company's other investment (Washington 1980).

In the first six years after the Washington statute was adopted, the three investor-owned electric utilities in Washington devoted about 16% of their rate base additions to investments eligible for the incentive (Blackmon 1991).

Kansas, Montana and Connecticut have also enacted statutes permitting utilities to earn bonus returns on DSM. In Kansas, the commission can award a bonus return of one-half to two percentage points on investments in conservation and renewables; commission staff believes, however, that such a bonus was never requested the first decade it was available (Sicilian 1990). In Montana, the return can be up to two percentage points greater than the company's ordinary return at the commission's discretion; utilities, however, have not requested the bonus (Eck 1988). The Connecticut statute allows the commission to set the return increment in a range from one to five percentage points. In its first application of the statute, the commission authorized a bonus of up to three percentage points for United Illuminating Company (Connecticut Department of Public Utility Control 1990).

In its 1986 WEPCO decision, the Wisconsin commission authorized an ROR increment tied to DSM performance. It allowed WEPCO to earn an additional one percentage point on the equity portion of its ratebased DSM for every 125 MW of demand reduction that WEPCO could demonstrate resulted from its programs (Wisconsin Public Service Commission 1986). WEPCO did claim a bonus under this clause, although there was controversy about the utility's estimates of demand reductions, an issue discussed in Chapter 11.

Amortization Period

A third dimension of ratebasing practice is the amortization period, during which expenditures are recovered and the return is earned on the unrecovered amount. The goal of providing efficient price signals would be served if the amortization period corresponded to the estimated service life of the investment. That number, however, can vary. Some DSM measures, such as compact fluorescent lamps, may last from two to six years, while other measures, such as electronic ballasts, are expected to last 15 years or more. Because useful lives can vary so greatly, a commission or utility proposing to ratebase DSM must decide whether it will:

• Estimate the life of each DSM measure and amortize it accordingly;

• Assign DSM investments to one of a limited number of expected lifetime groups (say, 3, 5, 10, and 15 years); or,

• Apply a blanket amortization period to all ratebased DSM investments.

Besides useful life, another factor in setting the amortization period is the utility's perception of the risk in ratebasing. If there is a possibility that cost recovery of ratebased amounts might be curtailed, then the utility may prefer a shorter amortization period, since the implied risk would be less. In fact, utility executives are often concerned that ratebased expenditures will not be fully recovered, since regulators can reverse their predecessors' decisions.

A single amortization period has usually been adopted for all rate-based DSM. At Puget Power, for example, DSM has been ratebased over ten years, although the statute establishing the bonus return specifies that the amortization period can be as long as 30 years. A ten-year period was also specified by the Wisconsin commission for WEPCO; the commission felt this was a reasonable estimate of the average useful life of the DSM measures that would be installed in WEPCO's programs (Wisconsin Public Service Commission 1986).

Responding to the perceived risk of nonrecovery, some utilities have suggested that amortization periods should be shorter than the expected useful life, at least until experience demonstrates that full recovery of ratebased investments is the norm. Jersey Central Power & Light, for example, suggested in a generic proceeding on cost recovery that the amortization period should be capped at five years (Raber 1991).

Financial Analysis

This section considers the financial effects of ratebasing on a utility and its customers. The first subsection below looks at the differences in utility revenue requirements under expensing and ratebasing. The second subsection analyzes the impacts of ratebasing on the utility's shareholders, as measured by cash flow. For these analyses it is assumed that the allowed return on ratebased DSM is identical to the firm's cost of capital and the return allowed on other elements of the rate base.

Impact on Revenue Requirements

Suppose a utility undertakes a one-year DSM program of a given size. Compared to expensing the program, what impact would ratebasing the DSM have on the utility's revenue requirements?

First consider the effects of ratebasing on *nominal* revenue requirements. (Calculations done in nominal dollars ignore the time value of money.) Also assume, for simplicity, the absence of corporate income taxes. A simple analysis shows that, relative to expensing, ratebasing would require collecting greater total revenues from ratepayers than expensing. The reason is clear: ratebasing provides full recovery of the original expenditure, *plus* an annual return on the unrecovered balance. Just the original expenditure is collected under expensing.

Figure 5-1 illustrates this simple case, assuming a one-time DSM

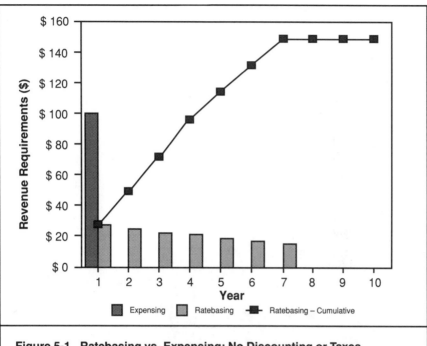

Figure 5-1. Ratebasing vs. Expensing: No Discounting or Taxes

program outlay of $100. The tall, dark bar illustrates the impact on revenue requirements of expensing: the $100 is recovered from ratepayers the year of the outlay. The series of shorter, shaded bars illustrates ratebasing with a seven-year amortization period and a 12% ROR applied each year to the diminishing unrecovered balance.[1] Finally, the line shows the cumulative effect on revenue requirements under ratebasing. Note that by year 5, the cumulative amount collected from ratepayers has exceeded the $100 that would have been collected under expensing, and that by the end of year 7 the total collected under ratebasing reaches $148.

Financial decisions, however, are customarily made on a present-value basis, i.e., adjusted for the time value of money. Again suppose there are no corporate taxes, and the discount rate equals the utility's cost of capital. In this simplified case, discounting would exactly offset

---

[1] The 12% figure is the assumed overall cost of capital and authorized return, based on an assumed 50/50 debt/equity capitalization, with debt and equity costs of 10% and 14%, respectively.

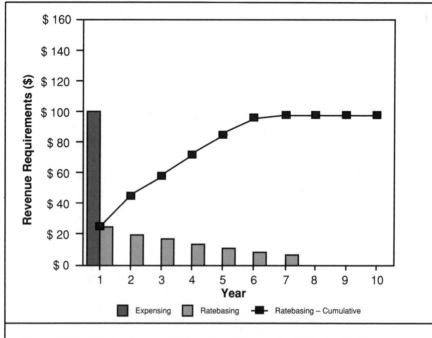

Figure 5-2. Ratebasing vs. Expensing: Discounted Values, No Taxes

the return component that is added to revenue requirements each year. The present value of the entire stream would be equal to the original program expenditure; therefore, ratebasing and expensing would have the same present-value effect on revenue requirements.

In Figure 5-2, all values have been converted to present values. The line tracking cumulative present value shows that when all seven years are considered, the present value equals $100, the same as the revenue requirement under expensing.

Next, add corporate income taxes to the analysis. Under ratebasing, the utility's DSM program expenditures would be financed by a combination of debt and equity capital. Before providing a return to equity shareholders (either as dividends or retained earnings), the utility would have to pay taxes on its pre-tax earnings. Therefore, each year's revenue requirements would be increased by the amount needed to cover taxes on the return to equity. No additional taxes would be paid, in contrast, if the program were expensed. Due to the tax effect, the present value of revenue requirements would be higher for ratebas-

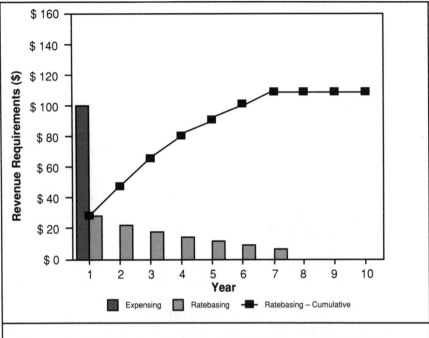

Figure 5-3. Ratebasing vs. Expensing: Discounted Values, with Taxes

ing than expensing, assuming again that the discount rate is the utili-ty's cost of capital. This point is illustrated in Figure 5-3, which assumes a corporate income tax rate of 34%. After discounting, the present value of the revenue requirements under ratebasing is $110, or 10% higher than under expensing.[2]

The preceding analysis follows industry practice in using the util-ity's cost of capital as the discount rate. But whether this is the correct rate to use in analyzing revenue requirements *from the ratepayer per-spective* is arguable, because the utility does not pay revenue require-ments; ratepayers do.

If ratepayers' discount rates are sufficiently greater than the utili-ty's cost of capital, the significance of the tax effect is diminished. Ratepayers with sufficiently high discount rates should prefer ratebas-ing to expensing. For a particular revenue requirements scenario, it is

---

[2] The 10% difference is illustrative. The actual difference will depend on several fac-tors, including the utility's capitalization ratio, the costs of equity and debt, and the tax rate.

possible to compute the crossover point, the discount rate at which ratepayers would be indifferent about expensing or ratebasing. For the scenario shown in Figure 5-3, the crossover point is 15.6%; if ratepayers' discount rate is greater, they should prefer ratebasing.

Ratepayers' discount rates may be higher or lower than the crossover point, and are almost certainly not homogeneous. An individual ratepayer who pays 19% interest on a credit card surely has a different discount rate than an institutional ratepayer such as a municipal government that can borrow money at tax-exempt rates. Since there is no single discount rate that correctly represents all ratepayer preferences, blanket statements that ratepayers will prefer ratebasing to expensing, or vice versa, should be viewed skeptically. It is valid to conclude, however, that for discount rates in the vicinity of the utility's cost of capital, the choice between expensing and ratebasing will not result in *substantial* differences in the present value of revenue requirements.

Impact on Utility Cash Flow

From the standpoint of financial theory, the utility's financial objective should be to maximize shareholder wealth, as measured by the net present value of cash flow accruing to shareholders in the form of dividends and retained earnings. How ratebasing affects cash flow should determine its value to the utility.

Again suppose a utility undertakes a one-year DSM program of a given size. Compared to expensing the program, how would ratebasing affect the utility's cash flow?

For this analysis, assume the utility operates in an idealized regulatory model: the regulatory commission sets the allowed rate of return on rate base equal to the utility's actual cost of capital; the test year used for ratemaking purposes correctly anticipates both sales and costs; and competitive forces do not preclude the utility from earning its allowed rate of return.

First consider expensing. The utility's revenues attributable to the DSM program are exactly equal to its expenditures, and the impact on cash flow is zero.

Now assume instead that the utility ratebases its investment in DSM and is compensated fairly for the cost of capital used to support the program, consistent with the idealized model. During the amortization period, the utility's cash inflows would be increased annually by the return and amortization components of the ratebased investment, plus an amount for taxes due on the portion of return attributable to equity. After taxes are paid, the residual cash flow is simply the return and amortization on the ratebased investment.

In nominal dollars, the net cash flow over the entire period (i.e., the residual inflows during the amortization period, less the initial outlay) would be positive. But when the inflows are discounted back at the utility's cost of capital, which is the correct discount rate for the utility perspective, their present value would exactly equal the initial investment in DSM. Thus, the net present value of the entire series of cash flows would be zero—the same as with expensing.

From the shareholders' perspective, this is at best a neutral result. While cash returns are increased, the value of these returns is fully offset by discounting for the time value of money. Therefore, ratebasing at the utility's cost of capital does not provide any real financial gain to the utility or its shareholders, even in an idealized regulatory world; ratebasing of this type does not offer any additional incentive to undertake DSM.

A corollary to this result is that real financial gain can occur only if the utility realizes a rate of return on ratebased assets greater than the cost of capital. How much greater a return would represent a true gain would be related to investors' perceptions of risk.

Risks and Benefits of Ratebasing as Perceived by Utilities

The previous analyses were based on a simple, idealized model of financial decision making in a regulatory environment. In the real world, however, perceptions of ratebasing are usually colored by other issues. This section discusses the risks and benefits that utility managers and shareholders may perceive in ratebasing.

Risks

Incomplete Cost Recovery Due to a Changing Regulatory Environment. Utility managers are frequently concerned that the regulatory commission will deny full cost recovery on ratebased DSM. A commission could exclude a part of the DSM expenditure from the rate base, or reduce or eliminate the allowed rate of return.

Why would a commission deny full recovery? This could conceivably occur if DSM programs failed to achieve their projected penetrations or to provide the forecasted impacts, or were less cost-effective than originally estimated. It also might happen if the decision to undertake the programs was judged imprudent, or if the programs were imprudently managed. Or, it could simply be the result of a change in the membership or philosophy of the commission.

Given the current regulatory environment in which DSM programs are frequently established by commission directive, the risk of curtailed cost recovery appears limited. Mechanisms in several states for preapproval of DSM programs further mitigate this risk. Nonetheless, it is a consistent theme in utility managers' comments on ratebasing and certainly affects their views.

Concern over full cost recovery also gives rise to utilities' preference for applying short amortization periods to ratebased DSM. A shorter period presumably reduces the risk that the full amount invested will not be recovered.

Incomplete Cost Recovery Due to Competitive Discounting. A related concern about incomplete cost recovery stems from growing competition in the utility industry. Competitive pressures from cogeneration and alternative fuels result, sometimes, in discounts for large customers. Until rates are adjusted in a general rate case, such discounting may effectively preclude the utility from collecting its full authorized revenue, including the amortization and return on ratebased DSM from previous years.

Effects on Competitiveness. As noted previously, total revenue requirements needed to recover an expenditure are greater for ratebasing than for expensing. Average rates must be higher under ratebasing, making the ratebasing utility less competitive vis-à-vis other fuels and increasing the risk of major customer loss.

Balance Sheet Risk. Many DSM expenditures do not create hard assets that are owned and controlled by the utility. Although the ratebased DSM expenditure is carried on the utility's books as an asset, it may be less secure, from the standpoint of investors, than some other utility assets. The value of a ratebased DSM asset is based on the expectation that regulators will allow future recovery of direct costs and a return on investment; as previously noted, utility managers believe there is some nonzero probability that full recovery will not be allowed. Further, in the event of extreme financial difficulties or bankruptcy, hard assets such as generating plants would likely be convertible to cash, whereas the value of ratebased DSM would be contingent on regulators' willingness to allow the sale of a regulatory asset.

Another problem with ratebased DSM is that it is not bondable property; it cannot be pledged as collateral to support utility debt issues. If ratebased DSM becomes a major component of a utility's balance sheet, the utility could conceivably reach the maximum level of unbondable property allowed by debt underwriters, in which case its ability to issue new debt would be constrained. Even if the limit were not reached, the utility's debt rating might decline.

Benefits

Maintaining or Growing the Rate Base. The previous theory-based analysis showed that ratebasing at the cost of capital does not yield a real financial gain as measured by cash flow. It follows that growth in the rate base per se has no value for utility shareholders unless they can expect to earn more than the cost of capital. Shareholder wealth is maximized by achieving high returns on investment, not necessarily by adding to the base on which the return is earned.

There may be other factors besides cash flow, however, that motivate utility management. Anecdotal evidence indicates that many utility managers believe growth in the rate base is desirable even if it occurs through investments that earn no more than the cost of capital. The desire for such growth may arise from erroneous information about how Wall Street values stocks, from a belief that managers in larger firms command higher salaries, or simply from the notion that "bigger is better." Also, many utilities, after substantial growth in rate base in the 1970s and 1980s due to major generating plant additions, now have adequate capacity and increasingly look to non–rate base sources such as independent power producers to meet future growth in demand. In such cases, adding DSM to rate base may cushion the change from the era of high growth.

DSM as a Profit Center. Even when ratebased DSM is earning no more than the utility's overall cost of capital, utility managers tend to speak of DSM as contributing to utility profitability; expressions such as "DSM is now a profit center" are common. This may enhance the status of DSM personnel within the firm and lead to a greater sense of job satisfaction.

Rate Stability. It was previously noted that the greater revenue requirements that follow from ratebasing might adversely affect the utility's competitiveness. In some circumstances, however, ratebasing could be considered preferable to expensing for competitive reasons. For example, a utility that is directed to undertake a large-scale DSM program over a short period (say, three years) might choose ratebasing over expensing because the latter would have a more pronounced impact on near-term rates and might result in a sudden surge in customers' interest in competing fuels. By amortizing program costs over several years, ratebasing can contribute to rate stability.

Empirical Results

As previously noted, in the 1980s several states authorized ratebasing of DSM and an associated bonus rate of return. To the author's knowledge, only one published study has attempted to assess retrospectively whether the ratebasing-with-bonus formula effectively stimulated DSM (Blackmon 1991). This analysis, drawn from Washington State's experience, found that utilities did not significantly shift spending from traditional supply-side options to bonus-eligible projects. Since the incremental investment was small, the effect of the incentive on rates was also small, less than 0.1%.

The analysis also pointed out a weakness of Washington's mechanism: between rate cases, utilities could not earn a return on DSM investment not already in the rate base. Such an allowance would be analogous to an AFUDC (allowance for funds used during construction) mechanism, which allows utilities to recover their carrying costs for new plants that are not yet in rate base. In the case of Puget Sound Power & Light, which had a four-year interval between rate cases in the late 1980s, it was estimated that the decision to invest in DSM in lieu of supply-side measures cost the firm $650,000 per year in earnings, notwithstanding the bonus rate of return that was eventually authorized.

Evaluation and Conclusions

Ratebasing is a means to recover the costs of DSM programs and, in some circumstances, a means to stimulate utility-sponsored DSM. Conclusions about its usefulness as a mechanism depend on the purposes for which it is considered.

Ratebasing as a Cost-Recovery Mechanism

Ratebasing of DSM expenditures that provide long-lived benefits is readily understandable, intuitively appealing, and theoretically sound. Placing DSM expenditures in rate base:

- Is consistent with the axiom of integrated resource planning that all resource options should be considered on an equal footing.

- Provides more efficient price signals than expensing, since it amortizes the costs of DSM over the period when benefits are received.

- Promotes intergenerational equity.

While DSM ratebasing is not an established practice in most states, it is easily accommodated within conventional accounting and regulatory policies. Since 1980, several states have authorized ratebasing. Administration of ratebasing is straightforward and yields predictable financial outcomes.

Some utilities may be attracted to ratebasing of DSM to compensate for declines in the book value of conventional supply-side investments. Utilities may also value ratebasing's ability to spread the rate impact of a large-scale DSM program over several years.

The main problem with ratebasing as a cost-recovery mechanism is that it creates additional risks in the minds of utility managers. The fact that it delays receipt of funds that might be immediately collected (i.e., expensed) raises concerns that full cost recovery may not be available for regulatory or competitive reasons. These concerns figure prominently in utilities' evaluation of ratebasing.

Given these concerns, is ratebasing a worthwhile cost recovery mechanism? If the DSM programs are so substantial that expensing them would significantly distort current rates, the answer is yes. In other circumstances, the risks and additional administrative and accounting costs may outweigh the benefits.

Ratebasing as a DSM Incentive Mechanism

In the appropriate circumstances, ratebasing can provide some bonus or positive incentive for DSM. It does not, however, by itself adjust for lost revenues or decouple utility earnings from sales.

Analysis shows that ordinary ratebasing—with the rate of return equal to the utility's cost of capital—does not provide the utility a bonus above costs. Any stimulative effect of ordinary ratebasing on a utility's DSM expenditures would be attributable to a desire for growth in rate base. According to financial theory, growth per se is not an appropriate goal for utilities; nonetheless, among utility managers growth in rate base may be valued.

Ratebasing can be expected to have a true stimulative effect on DSM only when the authorized rate of return on DSM in rate base exceeds the utility's overall cost of capital. How much greater the return has to be is debatable; authorized bonuses presently range from one to five percentage points additional return on the portion financed by equity. The two-percentage-point bonus return allowed on ratebased DSM in Washington State was apparently ineffectual as an incentive mechanism.

On balance, ratebasing is a weak incentive mechanism. Perhaps its greatest limitation is that it does not intrinsically create a connection between the utility's DSM performance and its financial reward (although it is possible to vary the size of the bonus according to some measure of performance). Instead, ratebasing rewards investment, raising the possibility that utilities will overinvest in ratebase-eligible items.

Acknowledgments

Portions of this chapter are based on an earlier report by the author, *Ratebasing of Utility Conservation and Load Management Programs*, published by the Alliance to Save Energy in 1988 under the sponsorship of Northeast Utilities. The author acknowledges the support provided by those organizations for his research on ratebasing.

References

Blackmon, Glenn. 1991. "Conservation Incentives: Evaluating the Washington State Experience." *Public Utilities Fortnightly* 127. January 15.

California Public Utilities Commission. 1990. Decision 90-08-068. August 29.

Chase, Bradford S. 1987. Quoted in Lucas Held, "More Control over New England Power Consortium Backed." *The Middletown [Conn.] Press.* July 7.

Connecticut Department of Public Utility Control. 1990. Order in Docket Nos. 89-08-11 and 89-08-12. January 24.

Eck, Eric. 1988. Personal communication with the author. Helena, Mont.: Montana Public Service Commission. September 14.

Idaho Public Utilities Commission. 1979. Order No. 14466. March 9.

Idaho Public Utilities Commission. 1980. Order No. 15861. September 26.

Krasniewski, R. J. and R. J. Murdock. 1980. "Residential Conservation Service Programs and Their Cost Recovery." *Public Utilities Fortnightly* 105. June 5.

Krzos, Joseph. 1988. Personal communication with the author. Madison, Wisc.: Madison Gas & Electric. December 16.

Maine. 1987. P. L. 1987, Ch. 613.

Massachusetts Department of Public Utilities. 1988. Order 86-36-F. November 30.

Raber, Marvin. 1991. Comments of Jersey Central Power & Light in Docket EX90040304. January 16.

Sicilian, Shirley. 1990. Personal communication with the author. Topeka: Kansas Corporation Commission. September 27.

Washington. 1980. Rev. Code Wash. 80.28.025.

Wisconsin Public Service Commission. 1986. Findings of Fact and Order in Docket 6630-UR-100. December 30.

Chapter 6

Sharing the Savings to Promote Energy Efficiency

Joseph Eto, Alan Destribats and Donald Schultz

Introduction

Shared-savings incentives are a new way for regulated utilities to earn money by encouraging customer energy efficiency. The basic idea is that the benefits of cost-effective energy efficiency measures can be shared explicitly among the customers participating in the utility program, all utility ratepayers, and the utility. For the participating customers, electricity bills are lowered directly. For utility ratepayers, the costs of providing electric service are reduced compared to the utility doing nothing to improve customer energy efficiency. For the utility, a fraction of the net savings to all ratepayers is retained as earnings.[1]

Business activities eligible for shared savings remain under the jurisdiction of traditional state regulatory agencies, but the methods used to calculate earnings differ fundamentally from both those used in traditional ratemaking and those used by other regulatory incentive mechanisms for DSM. First, eligible utility demand-side activities must have positive net resource value, which is different from traditional regulatory tests for the prudence and usefulness of utility supply-side investments. Second, since the utility's earnings are a fraction of this net resource value, the relationship between the earnings from shared savings and the traditional fixed rate of return earned on rate base may be only coincidental. Third, unlike other financial incentives to utilities for DSM, the earnings from shared savings accrue in direct proportion to the net societal benefit of the demand-side activity, so

---

[1] Shared-savings incentives to reward utility DSM activities were first proposed in Wellinghoff (1988).

that shared savings may be able to harmonize the utilities' incentive to increase earnings with the societal goal of a least-cost energy system.

However, departures from traditional ways of regulating utilities have risks that utilities and their commissions must evaluate, including:

• Uncertainty about the cost and performance of demand-side resources;

• Uncertainty about the value of these resources as avoided supply-side resources;

• Utility perceptions of the certainty of earnings from demand-side activities relative to other earning opportunities; and conversely,

• Commissions' certainty about the amount and timing of utility outlays and earnings.

It generally seems appropriate to distinguish among the risks that utilities and their commissions can and cannot control. For example, fuel-adjustment clauses have the primary effect of shifting risks associated with fuel price volatility onto the ratepayer. The rationale is that fuel price volatility is beyond the utility's control. In the case of shared savings, however, no one yet knows the magnitude of these risks and, consequently, the appropriateness of existing rewards.

This chapter reviews progress in striking the balance between risk and reward for shared-saving incentives for utility demand-side programs. This chapter begins with a brief description of the origin of the shared-savings concept with energy service companies because this background highlights the role of state utility commissions in adjudicating the risks and rewards of delivering energy services. After defining the basic elements of shared-savings arrangements for utility demand-side resources, we review recent experience in New England for two operating subsidiaries of the New England Electric System (NEES), and in California for the Pacific Gas and Electric Company (PG&E) and the San Diego Gas and Electric Company (SDG&E), comparing and contrasting specific details of the arrangements approved for each utility. We comment on the collaborative processes that led to the development of the incentives because they were instrumental for reaching consensus on the principle of providing positive earning opportunities to utilities for their demand-side activities and because they played a major role in the design of programs eligible for these earnings. Early financial results from the programs are then presented. In the final section, findings are summarized.

Origin of the Shared-Savings Concept for Utility DSM Activities

In the late 1970s, well-documented social and institutional barriers hindering the deployment of cost-effective demand-side resources (Blumstein et. al. 1980) created market opportunities for a new type of business dedicated to providing energy services, rather than energy forms per se (Sant 1980). Energy service companies (ESCos) acted as third-party developers, financiers, and in some cases operators of energy-efficiency investments on behalf of building owners or industrial firms that were unable or unwilling to pursue efficiency opportunities on their own. In return, ESCos retained a portion of the utility bill savings that resulted from their energy saving services. The agreements between ESCos and building owners came to be known as shared-savings agreements because the ESCos' earnings were directly related to the amount of energy they were able to save for a client.[2]

The experience of the ESCo industry during the past ten years is currently relevant for two reasons. First, the ESCo industry has tapped only a limited amount of the available, cost-effective, demand-side resource. The existence of these untapped resources has induced commissions to provide incentives to utilities to acquire these resources. Second, one of the most important reasons ESCOs have been unable to fully tap demand-side resources is that measuring energy savings is a formidable task, a major challenge for commissions and utilities when designing equitable shared-savings incentives.

What is new for utility shared savings is that the regulator, in effect, acts as an independent arbiter of energy savings. That is, the measurement dispute is no longer strictly an issue between an ESCo, or any energy service provider, and the client. Energy savings will become a central topic for the utility and its regulator because the regulator must allocate the risk of demand-side resource performance and value between the utility and its ratepayers. In this capacity, commissions must make the same type of determination that they make in determining the value of supply-side resource investments (Wiel 1990). The important difference is, that because energy savings can never be observed directly, there will always be an element of controversy. As we shall see, there is no standard to allocate this risk; no one has yet developed a precise prescription.

---

[2] The ESCo concept is fully described in Chapter 9.

Shared Savings Defined for Utility Demand-Side Programs

The basis for most utility shared-savings programs can be characterized using this simple formula:

$$NRV = (LR \times AC) - PC$$

where:

NRV = net resource value ($)
LR = load reductions (kW or kWh)
AC = utility avoided supply costs ($/kW or $/kWh)
PC = energy efficiency program costs ($), including utility administration, rebates, and customer contribution[3]

The basic idea is that when a utility invests in a cost-effective demand-side program, the program has a positive net resource value. In shared savings, this positive value is shared between the utility and its ratepayers. The utility's share is typically specified as a fixed percent of the net resource value (e.g., 10%, 13.5%, and 15% for NEES, SDG&E, and PG&E, respectively).[4]

As a result of this direct link between the net resource value of a demand-side investment and a utility's earnings, shared-savings incentives reward successful utility acquisition of cost-effective demand-side resources, rather than utility spending on DSM programs.[5] In this respect, shared savings differ fundamentally from most of the other incentive mechanisms described in this book because an explicit determination of net benefits, including energy savings, must be made.

Despite the simplicity of the concept, there are a variety of ways the terms in the equation can be defined, the incentive to the utility calculated, and qualifying utility performance measured.

Comparing Utility Shared-Savings Incentives

As noted in Chapter 2, shared-savings incentives have been approved in 13 states. However, shared savings are a new earnings opportunity

---

[3] There are subtle, but important, differences in the definition of program costs between various utility shared-savings incentives.

[4] In the case of NEES, in addition to 10% of *net* resource value, they also receive 5% of *gross* resource value—as described later in this chapter.

[5] Spending levels are the basis for some of the incentives considered in this book, such as ratebasing (see Chapter 5).

for utilities; the first shared-savings incentive was approved in 1989. Existing incentives are probably best regarded as experiments in progress. In other words, we fully expect that features of the incentives will undoubtedly change, perhaps dramatically, as commissions and utilities gain experience.

In this section, the shared-savings incentives approved for two of the operating subsidiaries of the New England Electric System, Narragansett Electric (NE) in Rhode Island, and Granite State Electric (GSE) in New Hampshire, are reviewed.[6] The shared-savings incentive approved for Narragansett Electric was the first of its kind in the United States. The shared-savings incentives approved for the Pacific Gas and Electric Company and the San Diego Gas and Electric Company, both in California are also reviewed.[7] Where relevant, selected features of the shared-savings incentives that have been approved for New York State utilities are described, although our descriptions are not intended to be comprehensive.[8]

This review of shared-savings incentives is organized around the following ten program features:

• Earnings calculation

• Determination of load reductions

• Determination of avoided costs

• Determination of program costs

• Program cost recovery

• Incentive recovery

• Performance thresholds

• DSM program spending and shareholder earning caps

• Program eligibility

• Treatment of lost revenues

Before beginning, we would like to point to two general considerations. The size of a utility, its energy efficiency programs, and asso-

---

[6] The largest operating subsidiary of NEES is Massachusetts Electric (ME), which has a bonus-type incentive for its DSM activities. Due to ME's size relative to NE and GSE, certain aspects of its DSM activities are mentioned that are directly relevant for the incentives earned by NE and GSE.

[7] The review of programs for the California utilities is based on previous work by Schultz and Eto (1990).

[8] For a detailed discussion of the shared-savings incentives of New York utilities, see a recent summary by Gallagher (1991).

Table 6-1. Comparison of Utilities

	PG&E	SDG&E	NEES	
			NE	GSE
Electric revenue (B$)	5.9	1.2	0.4	0.04
Sales (BkWh)	68.2	13.4	4.5	0.60
Customers (M)	4.1	1.1	0.3	0.03
Average revenue (¢/kWh)	8.7	8.6	7.8	7.4
Source: Energy Information Administration, 1991.				

ciated regulatory staff have a tremendous influence on the formulation of the incentives discussed in this chapter. These differences are alluded to in Table 6-1, which compares utilities by electricity revenue, sales, customers, and average revenue per kilowatt hour. Particular features of the California shared-savings incentives are described in greater detail, due to the size of these efforts. The viability of California-style shared-savings incentives in states with smaller utilities and commissions is clearly a legitimate concern.

Second, even though we discuss program features separately, these features are interdependent, and thus it is extremely important to evaluate program features in aggregate to understand their net impact and how they counter-balance one another. Our findings are summarized in Table 6-2.

Earnings Calculation

While the "share" of the savings available for the utility to earn is relatively simple to determine, the actual financial benefits to the utility are complicated by various definitions and conventions associated with their calculation, chiefly the definition of demand-side resource costs and timing for the recovery of the incentive, both of which are discussed later in this chapter.

Of greater importance for the present discussion is the hybrid nature of the shared-savings incentives approved for the two NEES subsidiaries. The NE and GSE incentive programs involve, in addition to a share of the savings, a "maximizing incentive" that scales directly with the total value of avoided resource savings (i.e., before subtracting program costs). For these utilities, the net benefit of shared savings programs is a combination of a share of the net resource savings and a further incentive to aggressively pursue all DSM opportunities.

Table 6-2. Comparison of Utility Shared-Savings Incentive Programs				
	New England Electric System			
Program Category	**Narragansett Electric Rhode Island**	**Granite State Electric New Hampshire**	**Pacific Gas & Electric California**	**San Diego Gas & Electric California**
Utility earnings	10% of NRV plus 5% of avoided cost benefit (see Table 6-3 for sample calculation)		15% of NRV (see below for definition of program costs)	13.5% of NRV
Energy savings	Participation based on utility records; per-participant savings based on engineering estimates that are updated for future year programs using detailed program evaluations of current year programs			
Avoided costs	Determined by NEES system planners annually for life of current year program		Set annually (for life of current year program) in pre-existing proceedings to determine long-run marginal costs	
Program cost	Includes both utility and customer costs; utility cost based on company records; customer contribution estimated		Includes only utility costs, based on company records	Identical to NE
Program cost recovery	Expensed annually			
Incentive recovery	Life-cycle program benefits fully recovered in year following program start		Life-cycle program benefits recovered over 3 years from program start	
Performance threshold	No earnings on first 50% of estimated overall savings, but no penalty	Earnings on all savings, provided 50% threshold is exceeded; no penalties	Program-by-program participation targets trigger receipt of incentives or, for sub-par performance, penalties (see Table 6-4)	
Earnings or spending caps	None, differences in overall expenditures of greater than 10% must be reported quarterly		Spending cap of +30% of authorized budget. Earnings cap of +10% of pre-program estimate	
Program eligibility	All demand-side activites treated as a package		Only demand-side activities explicitly designed to displace supply resources eligible; other demand-side activities subject to non-shared-savings incentives (see Table 6-6)	
Treatment of lost revenues	Annual FERC rate case for generating subsidiary, New England Power Co., reconciles revenues due to differences between forecast and actual sales		Electricity Revenue Adjustment Mechanism maintains balancing account to reconcile differences between forecast and actual base rate revenues on an annual basis; fuel adjustment clauses treat impacts on variable cost fluctuations	

A sample calculation of NE's incentive appears in Table 6-3. It only applies when the net resource benefits exceed a 50% threshold.[9] Note that evaluation and customer costs (lines 2 and 3 in Table 6-3) are subtracted from the total avoided utility supply costs prior to calculation of these thresholds. Finally, the maximizing incentive (line 10) is subtracted from the net benefit before the NE share is calculated. Table 6-3 illustrates that the maximizing incentive (line 10) approaches the size of the shared-savings incentive (line 11) and is an integral part of the overall incentive to the utility.

One rationale for the maximizing incentive is that it provides an earnings opportunity to the utility for demand-side activities that do not always have significant net resource benefits, such as some residential programs. Without this type of incentive, a profit-maximizing utility with limited budgets and staff will tend to pursue only the most cost-effective demand-side activities, usually in the commercial sector. In California, the issue of "cream-skimming" and the importance of utility delivery of demand-side programs aimed at other goals besides net resource value is addressed through performance thresholds and program eligibility.

As the example in Table 6-3 demonstrates, the percentages themselves are not particularly revealing without describing the mechanics for their calculation, the programs to which they apply, and the pre-existing ratemaking environment in which they are set. At this point in our review, we merely note that they range from 10% to 15%. In New York State, the incentives range from 5% to 20%, but other features of the incentives must be considered to make these levels comparable (Gallagher 1991).

Determination of Load Reductions

Measuring load reductions (either kW or kWh) is an imperfect science. In principle, load reductions can only be measured after a program or measure has been installed for some time. A particularly problematic issue is how to properly account for effects that are not within the control of the utility but that affect load reductions, such as weather or occupant behavior. Another issue is "free riders" or load-reducing actions that the customer would have undertaken anyway, even in the absence of the utility's program.[10] In this discussion, only two specific

---

[9] The use of performance thresholds is described later in this chapter. For NE, incentives are earned on all savings beyond 50% of overall program goals. For GSE, incentives are earned on all savings, not just those in excess of 50%, but only when the threshold has been exceeded.

[10] Evaluation issues are discussed in Chapter 10.

Table 6-3. Calculation of Maximizing and Efficiency Incentives (1990 M$)— Narragansett Electric Company

Line		1990 M$
1	Total avoided cost benefits	42.3
2	Evaluation costs	0.4
3	Customer direct costs	1.8
4	Total adjusted program value	40.1
5	Base value (50% of program goal)	13.7
6	Qualifying value (in excess of 50% threshold)	26.4
7	Utility program costs (not including evaluation or customer costs)	14.3
8	Base costs (50% of program goal)	4.9
9	Qualifying cost (in excess of 50% threshold)	9.4
10	Maximizing incentive (based on qualifying value)	1.3
11	Efficiency incentive	1.6
12	Total conservation incentive	2.9

Notes:
Line 4 = Line 1 − (Line 2 + Line 3)
Line 6 = Line 4 − Line 5
Line 8 = (Line 5 / Line 4) × Line 7
Line 9 = (Line 6 / Line 4) × Line 7
Line 10 = 5% × Line 6, but not less than zero
Line 11 = 10% × (Line 6 − Line 9 − Line 10), but not less than zero
Line 12 = Line 10 + Line 11

Source: Hutchinson, 1991.

issues related to calculation of utility earnings from shared-savings programs are discussed: (1) the separation of load reductions into two components—measuring participation in utility programs and measuring of load reductions per participant; and (2) the evolution of measuring load reduction per participant during subsequent program cycles.

The four utility shared-savings incentives discussed in this chapter distinguish two components of load reductions: (1) technology or measure performance; and (2) marketing or utility program performance. The first refers to load reductions per program participant, for which the utilities are not held directly responsible. The second refers to program participation, for which the utilities are held responsible.

Due to the accelerated nature of utilities' earnings from the shared-savings incentives (as discussed later in this chapter), estimates of load reductions per participant must be made before field measurements are available. Program participation, conversely, is determined from utility records. In other words, ratepayers bear the risk of a *measure's* demand-side performance on a per-unit basis, while the utility bears the risk of the performance of its demand-side *program*. This risk often translates to the level of participation obtained by the utility for its programs. However, the utility's risks are relatively modest since it influences the setting of program performance targets.

The estimates of a measure's performance, however, are not static. Because it is difficult to estimate a measure's performance, PG&E, SDG&E, and NEES (the parent of NE and GSE),[11] are comprehensively evaluating utility demand-side programs. The spending levels proposed by the utilities and management attention to program evaluation are expected to significantly advance the state of the art in this area. The outcome of these evaluations will be used to update the estimates of each measure's performance for future program planning. However, the revised per-participant/measure load reduction estimates can never retroactively reduce the savings figures per measure or participant that were used to develop incentive payments for the previous year's programs.[12]

A major contribution of the shared-savings incentives in California has been the rigorous discussion of measurement issues. For the first year of the programs, values were adopted for first-year load reductions, decay in savings over time, lifetimes, and free-rider fractions on a measure-by-measure basis. More importantly, acceptable techniques for evaluating and revising these variables over time were agreed on.

Determination of Avoided Costs

Load reductions are multiplied by utility avoided supply costs ($/kW and $/kWh) to obtain the total benefit of demand-side programs. The

---

[11] NEES's program evaluation will focus on ME's demand-side activities in Massachusetts, which has a bonus-type incentive based on measured evaluation results. Results from these evaluations will be used by both NE and GSE, after appropriate adjustments for conditions unique to each service territory.

[12] It is interesting to note that none of the shared-savings incentives allows the findings from the measure evaluations to update future year estimates for measures installed in prior years. This is partly due to the accelerated nature of incentive recovery. But more importantly, it is symbolic of the give and take involved in the negotiations that led to development of the incentives. In contrast, the non-shared-savings incentives to be earned by ME will be based on after-the-fact measurement of load reductions.

primary concern in establishing these costs for incentive determination purposes is ensuring that they are consistent with other utility uses of avoided costs. Without this consistency, the utility will have an incentive to manipulate these values to increase the apparent net resource value of the programs and consequently their earnings.

For the four utilities, treatment of avoided supply costs is similar to estimating energy savings per participant. To calculate net benefits from programs so that utility earnings can be quickly recovered, avoided supply costs are fixed for the programs' life. In NEES, these long-term values are determined annually by NEES's wholesale subsidiary, with review and approval by FERC. In California, they are established by an ongoing, pre-existing regulatory forum for resource planning.

Avoided supply costs are defined strictly in terms of direct costs avoided for all four utilities. In addition, other costs, external to the utility's direct costs, are avoided by reliance on demand-side rather than supply-side resources. One notable example of such external costs is the environmental damage caused by the construction and operation of supply-side resources (Ottinger et. al. 1990). In New York State, *dollar* estimates of these values are being included to determine the *avoided cost benefit* of demand-side resources eligible for shared-savings incentives (Gallagher 1991).

Another increasingly significant avoided cost is avoided transmission and distribution (T&D) facilities. The avoided costs used in the PG&E and NEES shared-savings incentives explicitly include these costs. A general concern when avoided T&D costs are included is that the programs eligible for these incentives must be targeted to locations that, in fact, have avoidable T&D facilities (Rosenblum and Eto 1986).

Determination of Program Costs

The societal cost of demand-side resources includes the utility's and the customer's expenses. If both are included and netted out from the benefits of avoided utility supply costs, the shared-savings formula is similar to the total resource cost test. If only the utility's costs are included, the formula becomes similar to the utility cost test.[13] Both approaches are used to determine shared-savings incentives.

Both NEES subsidiaries (NE and GSE) and SDG&E include customer and utility costs in the calculating their shared-savings incentives. PG&E includes only utility costs. There are good reasons to

---

[13] See CPUC/CEC (1987) or Krause and Eto (1988) for a formal definition of these cost-benefit tests for demand-side resources.

support either choice. On the one hand, inclusion of customer costs is truer to the total resource cost standard. On the other hand, customer costs (like energy savings) are difficult to measure, and, in any case, utility incentives to minimize its costs to deliver demand-side programs by reducing the incentives paid to participating customers are stronger if they are not combined with customer costs. Consideration of only utility costs will, however, tend to make the utility's "share" of the savings larger relative to a share based on the difference between avoided supply costs and the combination of utility and customer costs. In addition, when incentives are based only on utility costs, societal costs (utility costs plus customer costs) may increase since these costs are of minor concern to the utility.

In a practical sense, the importance of these definitions depends on specific DSM program designs. For example, the DSM programs of the NEES subsidiaries usually pay most of the demand-side resource cost. The customer contribution is nearly zero. In this case, utility cost and total resource cost tests would yield essentially the same result.

When customer costs are included in the calculation of shared-savings incentives (NEES and SDG&E), determination of customer costs is analogous to that for energy savings. In both cases, per-unit estimates are agreed on in advance because it is difficult to measure actual customer costs. In addition, information on customer costs is collected for updating the estimates that will be used in future year's programs. These estimates will not retroactively affect earnings from previous programs.

For PG&E, when customer costs are not included, each program must first pass the total resource cost test, which does include customer costs. As with the incentives for NEES and SDG&E, the customer costs used in the total resource cost test are estimates that will be updated for future programs.

Program Cost Recovery

One of the most important features of the four utility shared-savings programs involves timely recovery of program costs. Utilities' uncertainty about regulators' treatment of these costs has been cited as a major barrier to utility participation in demand-side markets (Chamberlin and Hanser 1991). All four utility programs provide immediate recovery of program costs as operating expenses in the year they are incurred. Expensing demand-side program costs has gone a long way toward increasing each utility's comfort with acquiring demand-side resources. An alternative to expensing, ratebasing, is described in Chapter 5.

Incentive Recovery

The net resource benefits from demand-side activities accrue annually for the life of the measure. However, the shared-savings incentives earned by NEES, PG&E, and SDG&E are recovered in advance of the useful life of the measures. For both NEES subsidiaries, the utilities' share of the entire life-cycle benefits from a given year's activities are recovered in full by the end of the year after those activities are verified. For PG&E and SDG&E, benefits are also accelerated, but they are spread over the first three years following program delivery.

Both the utilities and their commissions have reasons to accelerate the shared-savings incentive. From the utilities' perspective, delayed earning of shared-savings incentives increases the risk that the earnings will not be recovered because of, among other things, changing regulatory philosophies. Commissions, too, cite reasons for wishing to accelerate utility shared-savings earnings. First, accelerated earnings increase certainty about the total amount of ratepayer dollars to be paid. Second, accelerated earnings increase the visibility of the profits from demand-side activities to the utility. This reason is also shared by utility demand-side program managers. Third, accounting is simplified when multiple program elements, each with a different lifetime, do not need to be tracked separately.

Accelerated incentive recovery is similar to front-loading payments to qualified facilities (QFs) in power sales agreements with utilities. In California front-loading became controversial because of a perceived oversupply of QF power in the mid-1980s and, as a result of falling real (net of inflation) oil prices, charges that QFs were being overpaid. In the present context, these concerns are largely addressed by (1) the need for eligible programs to pass the total resource cost test, and (2) spending caps that, in California, limit the maximum level of activities on an annual basis or, in New England, trigger regulatory review when budgets are exceeded. Conversely, because incentive recovery is guaranteed, ratepayers have no recourse if subsequent evaluations reveal that performance has fallen short of expectations. For front-loaded QF contracts, substantial penalties are levied for substandard performance. For California utility shared-savings incentives, as discussed below, penalties are in place for sub-par program participation, but not sub-par measure performance.

Performance Thresholds

Performance thresholds, a central feature of the California shared-savings incentives, serve as regulatory sticks by specifying explicit earnings penalties if utility DSM program participation goals are not met. Performance thresholds are also present in the NEES shared-savings incentives, but they are specified in a more aggregate manner.

California performance criteria, designed to assess performance penalties for sub-par utility performance, were developed in response to utility underspending of authorized conservation and load management budgets during the mid to late 1980s (Caldwell and Cavanagh 1989). The effect is that the shared-savings earnings can be substantially reduced or even become operating losses if performance fails to meet expectations.

The California performance criteria are defined on a program-by-program basis. Performance is measured by program participation, not by program energy savings. This effectively separates the risk of the conservation measure's performance, which is deemed to be beyond the control of the utility from program participation, which is deemed to be within the control of the utility. There are three steps. First, an annual target level for program participation is set by the utility. Second, a minimum performance threshold or fraction of the target level is established. If participation fails to exceed this threshold, no incentives are earned. If participation exceeds the threshold, incentives are earned on the entire amount of net savings from the program. Third, a "deadband" is established below the minimum performance level. Penalties accrue if participation falls below the deadband.

The target levels and performance thresholds are set for individual programs (see Table 6-4). Both the goals and minimum performance criteria reflect the utility's and commission's confidence in the probability of program success. Mature programs may have high goals and minimum performance thresholds while goals for new or experimental programs may be defined more modestly. Since goals and thresholds are specified program by program, utility cream-skimming can be mitigated somewhat by establishing high goals and thresholds for less cost-effective (i.e., less profitable) programs, which might otherwise be neglected.

NEES's performance criteria are specified on an aggregate basis: incentives are only earned when energy savings exceed 50% of overall DSM program goals. This specification allows the utility considerable flexibility in two dimensions. First the utility may reallocate efforts among individual programs throughout the year.[14] Second, threshold

---

[14]   A quarterly filing with the commissions is required when program spending differs from agreed rates by more than 10%. These filings may then become the basis for subsequent regulatory intervention although this has not happened in New England.

Table 6-4. Minimum Performance Thresholds (% of Participation Targets)—Pacific Gas & Electric, 1991 program

Program Category	Minimum Performance Thresholds for Incentive Payments (% of participation goals)
Commercial, industrial, agricultural energy management incentives	75%
Commercial new construction	25%
Residential new construction	30%
Residential appliance efficiency	75%
Commercial, industrial, agricultural energy management services	70% 75% for commercial
Residential energy management services	80%
Super-efficient homes	70%

Note: The thresholds represent percentages of participation goals that must be exceeded for the utility to earn incentives; if exceeded, the incentives are earned on the total benefits from the programs, not just those in excess of the thresholds. If participation is less than the threshold value, the utility earns no incentive. If participation is significantly below the threshold value, penalties are applied.

Source: Pacific Gas and Electric Company, 1991.

performance can be met through any combination of participation and savings per participant. In short, the specificity inherent in the California incentives is replaced in New England by a bottom-line orientation.

The specification of the NE performance criteria complicates calculation of net program benefits because the criteria act as earnings thresholds (see, for example, Table 6-3). No incentives are earned on the first 50% of projected savings; incentives are only earned on savings *in excess* of the 50% threshold. This means that the first 50% of program accomplishments, and utility expenditures, assuming these expenditures vary in direct proportion to savings, do not produce any incentive. Conversely, assuming the program target is reached, earning 10% on 50% of the savings means that only 5% has been earned on the entire program. For GSE, once the 50% threshold is reached, 10% is earned on all net savings, including the savings required to reach the 50% threshold.

DSM Program Spending and Shareholder Earning Caps

Spending caps limit the maximum a utility can spend beyond its authorized DSM program budget. Shareholder earning caps limit the maximum incentive a utility can earn. Both are discussed in this section because spending is often directly related to earnings. That is, since the incentives are based on prior estimates for savings per participant or measure installed, spending caps become, de facto, earnings caps.

DSM program spending caps may superficially seem contradictory; if energy efficiency is such a good idea, shouldn't program expansion be encouraged? But, unlimited expansion of demand-side programs may not be warranted for several reasons. Theoretically, the cost-effectiveness of demand-side programs can diminish with program size as avoided costs decrease and the difficulty (i.e., cost) of recruiting participants increases on a per-unit basis. However, fixed avoided cost values are generally agreed on in advance for the purposes of calculating incentives, and thus changes in per-unit values due to quantity changes are usually not reflected in incentives formulas. In addition, as spending increases, the amount of money that must be collected from ratepayers also increases, sometimes causing rate increases. Some utilities and commissions try to keep these rate increases to modest levels each year by limiting program expansion. Furthermore, administrating greatly expanded programs may be difficult for the utility and its commission in the short run.[15] Finally, unlimited earnings from demand-side activities raise the more fundamental issue of what ought to be the appropriate basis for utility earnings. All three issues reflect the experimental nature of existing, shared-savings programs. Improved methods for dealing with mid-year adjustments in program spending and earnings will evolve as all parties deal with the programs.

The California shared-savings incentives contain explicit limits on expanded DSM program spending. The limits are set at 30% beyond authorized program budgets. Shareholder earnings are limited to no more than 10% above anticipated levels. The NEES programs do not contain explicit limits on program spending or earnings, but changes in spending more than 10% are reported quarterly and, as a result, may become the subject of regulatory review.

In New York State, the shared-savings incentives for some utilities

---

[15] Slower program growth rates will give the utility additional time to "fine-tune" its programs. These efforts can increase the cost-effectiveness of programs by allowing utilities to modify aspects of their program designs (e.g., lower rebate levels, more effective recruitment strategies).

put a cap on earnings by linking the size of the incentive that can be earned to that which could have been earned under traditional rate-of-return regulation (Gallagher 1991). In other words, an independent measure is used to limit earnings from shared-savings incentives, in this case by linking the shared-savings incentive to profits achievable under traditional utility regulation. While in New York, this measure is based solely on the utility's program costs, the California PUC has proposed establishing similar limitations based on total program (i.e., including customer) costs.

Program Eligibility

The program eligibility criteria for DSM incentives vary considerably between the California utilities and NEES. Within the New England states where NEES's subsidiaries operate, it has been felt that only exemplary utility DSM programs should be eligible for incentives. Largely for this reason, GSE's DSM programs were the only utility programs in New Hampshire initially allowed to earn incentives. This philosophy also explains why NE is only allowed to earn incentives on savings in excess of a threshold. In California, all major utilities are eligible to earn incentives. Utilities are allowed to earn incentives on all eligible DSM programs, but, as described previously, the programs must first exceed minimum participation goals.

A second and more important area of difference is in the types of DSM programs eligible for incentives. All of NE's and GSE's demand-side activities are treated in aggregate when incentives are determined; i.e., the shared-savings incentive is based on the total impact of all demand-side activities. There are, however, two important subtleties. First, each activity taken separately must pass the total resource cost test. Second, many activities not directly related to the delivery of energy savings such as measurement and evaluation, are included in calculating total program costs.

California, on the other hand, has adopted a much more disaggregated approach. Demand-side activities are first identified by demand-side categories, and only those activities falling into certain categories are eligible to earn incentives. The categories distinguish between programs that are primarily oriented toward displacing supply resources and those that are primarily oriented toward other goals, such as equity or customer service. In addition, measurement and evaluation activities are explicitly separated from individual programs and are not eligible for incentives. This is also the case for NEES. Table 6-5 summarizes California's categorization of demand-side activities with examples of eligible programs and the type of available incentive.

Table 6-5. Matching DSM Programs with Shareholder Incentives—Pacific Gas & Electric

Program Category	Examples	Incentive Treatment
Resource	Residential, commercial, industrial, and agricultural rebates; residential and commercial new construction	Shared Savings
Equity/ service	Direct assistance; residential, commercial, industrial, and agricultural audits; super-efficient homes pilot program	performance-based earnings adder
Other	Innovative rate design; measurement and evaluation; general administration	no incentives

Source: Pacific Gas and Electric Company, 1991.

California's approach recognizes that utilities have multiple reasons for intervening on the demand side. Shared savings, as an incentive for these activities, only make sense for activities with the primary objective of displacing supply resources. Other equally important demand-side activities should not be subject to the same incentive structure because the motivation for them is often legitimately quite different. These programs include those developed for equity considerations, such as certain residential programs. Similarly, demand-side activities with impacts that are difficult to measure, such as information or rate design programs, are probably also inappropriate for shared-savings incentives.

For demand-side programs that are primarily equity or service oriented, performance-based earnings adders were adopted. These adders are essentially cost-plus or bonus-type incentives that are triggered by achieving some measurable level of performance, such as number of audits provided.[16]

Treatment of Lost Revenues

A potentially complicating issue when comparing the net benefit of shared-savings incentives is the relationship between the earnings from shared savings and the sales revenue losses that are associated with utility demand-side interventions. Some say these losses should

---

[16] For NE and GSE, incentives for less cost-effective programs provided implicitly through the use of the "maximizing incentive" previously described. This feature allows the utility to earn incentives despite the low net resource value of certain demand-side activities.

be netted out from any calculation of the benefits of a shared-savings incentive. In fact, the issue is probably more philosophical than practical.[17]

California is well known for the Electricity Revenue Adjustment Mechanism, or ERAM, which establishes a balancing account to ensure that an approved revenue requirement is earned independent of sales volumes (see Chapter 3). It is less well known that New England Power, NEES's wholesale electric subsidiary, which collects 70% of NEES's revenues, has in effect a revenue adjustment mechanism on file with FERC. FERC annually approves New England Power's wholesale rates using a future test year. The result is that because the rates are determined annually and because demand-side activities are accounted for explicitly in the future-test-year forecasts, there is little room for unanticipated, "lost" revenues. Discrepancies, to the extent that they persist for any reason, including weather, business cycle, and DSM, are effectively "trued-up" in the following year's filing.

Thus, for all four utilities, demand-side activities that reduce sales beyond levels predicted in the rate-setting process are addressed by either explicit or implicit balancing accounts, which ensures that authorized revenue requirements will be earned. Uniform decoupling of revenues from sales for the four utilities facilitates comparisons among their shared-savings incentives, but it makes it difficult to transfer results to utilities in states where different ratemaking practices make "lost revenues" a more serious issue.

Evaluating Shared-Savings Incentives

In reviewing the calculation of utility earnings from shared-savings programs, it is apparent that the bottom line can only be determined by considering the combined impact of all incentive components. The utility's share of earnings can be increased either by providing an increased share (percentage) of the net resource benefits, by bonuses earned in addition to a percentage of the net savings, by using an avoided cost that includes externalities, or by excluding the customer's contribution from program costs. Conversely, earnings can be decreased by providing the utility with a lower share of the net resource benefits, by program thresholds below which no incentives are earned, or by the inclusion of programs whose cost-effectiveness may be low or indeterminate and indirect program expenses, such as measurement and evaluation in the overall package of programs eligible for incentives.

---

[17] The existence of "lost revenues" is really just a manifestation of the failure by traditional regulation to account for fluctuating sales volumes whatever their cause.

In this section, we attempt to assess these earnings trade-offs. Our discussion begins by describing the non-traditional regulatory settings from which the incentives arose because they provide important background information on the role of negotiations. Next, 1990 program results are used to assess quantitatively the profitability and significance of these utilities' demand-side activities.

The Role of Collaborative Negotiations in the Designing of Shared-Savings Incentives

The shared-savings incentives for the four utilities arose from "collaborative" negotiations that proceeded outside traditional regulatory forums. These negotiations were responsible for both the acceptance of the idea that it would be appropriate to reward utilities for their energy efficiency activities and for the specific incentive designs reviewed in the previous section. In particular, the informal setting of the collaborative process allowed for explicit bargaining and trading-off among various incentive design features. It is, therefore, misleading to evaluate the program design features reviewed in the last section in isolation. The combined effect of these features not only determines the financial bottom line, it also attempts to balance the risks and rewards inherent in the programs.

For example, all the incentives include minimum performance thresholds below which no earnings (and, in California, penalties) apply. This feature is designed partly to ensure a serious utility response to the incentives being offered. Concerns were expressed that, without these thresholds, no guarantees would ensure that utilities would aggressively pursue energy efficiency opportunities. In other words, the availability of financial incentives was predicated on a commitment by the utility to obtain significant savings.

In California, thresholds were also specified on a program-by-program basis to ensure that all customer groups would be able to participate in utility-sponsored energy efficiency activities. This feature, intended to limit utility cream-skimming in more lucrative energy efficiency markets, is a contrast to the bottom-line orientation of the NE and GSE shared-savings incentives whose thresholds are based on total program savings. In effect, the commissions and utilities must balance equity concerns against the need for flexibility with a relatively untested incentive. It is difficult to argue that one approach is superior to the other; in both cases, utility and commission staffing and priorities were different. Indeed, to the extent that in the future the balance is determined along with a host of other utility DSM policy issues, with or without a collaborative process, there may never be a conclusive answer.

Another example of the risk balancing reached through consensus in the collaborative is the decision to base first-year program savings per participant, including, in California, energy and peak demand savings, free-rider fractions, and persistence, on estimates that are now assumed to remain unchanged for the lifetime of the measures installed in the first-year programs. In effect, this decision transfers most of the risks of demand-side measure performance to the ratepayer. In return for protection from the performance risk of their demand-side activities, however, the utilities agreed to initiate large-scale evaluations of their programs to measure these risks precisely.

The design of the shared-savings incentives was the result of collaborative negotiations among stakeholders. While one can argue that the same results could have emerged from traditional regulatory forums, it is doubtful they could have emerged as quickly as they did in New England and California. In both cases, shared-savings incentives were established within one year after the initiation of discussions.

Initial Results from Utility Shared-Savings Incentive Programs

Table 6-6 presents 1990 program results for PG&E, SDG&E, and NEES. For the shared-savings portions of the utilities' demand-side activities, details are presented relevant to calculating the incentive, including the expected avoided utility supply costs (life-cycle energy and capacity savings times avoided costs), and the utility and customer costs, which when subtracted from the avoided costs yield the net resource value of the programs. The shared-savings and other incentives where applicable are reported. In addition to information specific to the utilities' shared-savings programs, summary information on aggregate demand-side activities and earnings is reported for PG&E, SDG&E, and NEES.

To evaluate the relative impact of shared-savings (and other DSM) incentives on utility operations, we use two crude ratios: (1) the percent of total utility operating revenue accounted for by demand-side programs in order to measure the role of demand-side activities in overall utility operations; and (2) the earnings resulting from incentives as a percent of utility demand-side program expenditures in order to gauge the profitability of demand-side activities.

We also present an indicator that measures the cost premium associated with shared-savings incentives by expressing the utility shared-savings earnings as a percent of the utility and customer costs for the program. This ratio measures the added cost to society and ratepayers

Table 6-6. Comparison of Utility Shared-Savings and Overall DSM Program Performance (1990 M$)

	PG&E		SDG&E		NE	GSE	NEES
	SS[1]	Total[2]	SS[1]	Total[2]	SS[1]	SS[1]	Total[2]
Avoided utility supply costs	115.4		21.7		42.3	4.4	
Utility DSM program expenditures	20.6	141.0	4.0	16.7	14.7	1.7	71.2
Estimated customer contribution	17.9		4.5		1.8	0.2	
Net resource value Total resource cost test Utility cost test	 94.8		13.1		25.8	2.5	
Shared-savings incentive	14.2		1.8		1.6	0.2	
Other incentives		1.6	0.2	8.0	1.3	0.2	5.0
Total incentive	14.2	15.8	2.0	10.0	2.9	0.4	8.3
DSM expenditures as a percent of utility revenues	0.2	1.5	0.2	0.8	0.6	0.1	3.8
Total incentive as a percent of DSM program expenditures	69	11[3]	50	60	20	24	12
Total incentive as a percent of utility program cost and customer contribution	37		24		18	21	

Notes:
[1]SS = shared savings
[2]Total = Total DSM program, including components eligible for shared-savings incentives
[3]PG&E's return of 11% on all 1990 DSM activities may be misleading because PG&E's incentive earning programs only began in the second half of 1990. A more proper measure, if data had been available, would be to express the earnings ($15.8 million) as a fraction of PG&E's spending on DSM in the second half of 1990, which was less than the $141.4 million spent over the entire year. In this case, the percentage earnings would be significantly larger.

Sources: PG&E, 1991; SDG&E, 1991; Hutchinson, 1991.

represented by the incentives to the utility. In other words, this ratio accounts for the way incentives, in effect, raise the cost of delivering energy efficiency.

We also present aggregate information for the utilities' entire DSM program. These numbers, presented under the heading "Total" in Table 6-6, include the utility shared-savings programs. The reasons for presenting aggregate results differ slightly for each utility.

Both PG&E and SDG&E sponsor demand-side activities that do

not receive shared-savings incentives. Some of these activities, however, are eligible for other incentives. More important, the receipt of shared-savings incentives for some demand-side activities is probably, in some sense, conditional on the utility's offering of these other, non-shared-savings activities. In other words, for PG&E and SDG&E, the shared-savings incentives must be viewed as one component of a utility's overall DSM activities.

Aggregate results for NEES are also appropriate because NEES has centralized program operations. Centralized planning, operation, and evaluation costs cannot be easily allocated to activities in individual service territories. For example, NEES's major program evaluation activities will take place in the Massachusetts Electric service territory. The costs will be borne by NEES and will consequently not show up on Massachusetts Electric's budget or on NE's or GSE's, yet these evaluation results will be used to determine savings and incentive earnings from future programs for all three operating companies.

The shared-savings components of California utility DSM programs are modest, accounting for no more than a quarter of total utility DSM activities.[18] However, for both PG&E and SDG&E, shared-savings programs were only in operation during the last half of 1990. They are approximately 50% of what they might have been if they had been operating for the entire year. Total DSM activities for the entire year, which include the shared-savings programs, account for measurable percentages of PG&E, SDG&E, and NEES operating revenue (1.5, 0.8, and 3.8%, respectively). NEES's DSM expenditures represent the largest percentage of operating revenue among the three utilities.

Shared savings appear to be profitable for the utilities. The shared-savings components of the utilities' demand-side activities produce earnings of up to nearly 70% (PG&E) on expenditures for utility DSM programs that are eligible for shared-savings incentives.[19] In general, both PG&E and SDG&E shared-savings incentives are more profitable (69 and 50%, respectively) than those of NE or GSE (18 and

[18] As previously noted, all NEES's DSM programs are rewarded with incentives, so this distinction cannot be made for NE, GSE, or NEES.

[19] It is tempting but not possible to compare these returns to authorized utility returns on undepreciated rate base, which are typically 11–13%. First, return on rate base is earned annually for the accounting life of the depreciating rate base; shared-savings incentives are earned on an accelerated basis either entirely in the first year (NEES) or over the first three years (PG&E and SDG&E) after the program has been established. Second, not all DSM program expenditures would be eligible for inclusion in rate base; only capital expenses are typically included in rate base. Third, and most important for PG&E and SDG&E, as mentioned previously, shared-savings program expenditures and incentives must be considered jointly with all of these utilities' DSM earnings.

21%, respectively) from the standpoint of return on shared-savings DSM program expenditures. Part of the reason is that NE only earns incentives on program savings in excess of a 50% threshold. More important, PG&E and SDG&E are engaged in many DSM activities that are not eligible for shared-savings incentives, while all of NEES's DSM activities (except measurement and evaluation) are considered in calculating incentive payments. Conversely, NEES's DSM incentives also include a maximizing incentive, which is not based on the shared-savings concept. These additional incentive features complicate direct comparison of the shared-savings components of the utility's DSM activities and highlight the appropriateness of examining all incentives jointly in the context of the utility's total DSM activities.

When shared-savings and other DSM incentive earnings are compared to all DSM activities, the overall returns for PG&E and NEES are more modest, 11 and 14%, respectively.[20] On the other hand, SDG&E's overall DSM program earnings are quite remarkable. SDG&E's non-shared-savings incentives are so profitable that the overall return on expenditures for their program (60%) is higher than the return on the shared-savings DSM activities. In fact, the returns were even higher initially, due to the absence of earnings caps on the non-shared-savings portion of SDG&E's programs.[21] As a result of this apparent oversight, SDG&E, in its filing to the CPUC for its incentive, claimed $6.2 million less than it would have otherwise been entitled to under the original terms of the non-shared-savings incentive. Even with the reduced claim for incentive earning, SDG&E's DSM programs are the most profitable of the three utilities.

Incentives represent an added cost to society for delivering energy efficiency. The shared-savings incentives paid to PG&E raise the total cost (customer costs plus utility program costs) of the shared-savings incentive-eligible demand-side measures to society by nearly 40%. For SDG&E, NE, and GSE, the cost premiums are more modest, ranging from 18 to 24%.[22] In part, these cost premiums reflect the high cost

---

[20] PG&E's return of 11% on all 1990 DSM activities may be misleading because PG&E's incentive earning programs only began in the second half of 1990. A better measure, if data had been available, would be to express the earnings ($15.8 million) as a fraction of PG&E's spending on DSM in second half of 1990, which was less than the $141.4 million spent over the entire year. In this case, the percentage earnings would be significantly larger.

[21] These programs were approved prior to the California Collaborative.

[22] Recall that these cost premiums reflect only the added cost of measures eligible for shared-savings incentives. For both PG&E and SDG&E, significant portions of the utilities' DSM activities are not eligible for shared-savings incentives, although they may be eligible for other, non-shared-savings incentives.

effectiveness of the DSM activities; all the programs continue to pass the total resource cost test with the inclusion of the incentives. More importantly, they reflect the limited experience of both commissions and utilities in determining what is the appropriate level of incentive for utility delivery of customer energy efficiency programs. It is clear, however, that the incentives paid to these utilities have added measurably to the cost of delivering energy efficiency.

Summary

Shared savings can provide positive incentives to utilities for DSM. In the examples we reviewed in California (PG&E and SDG&E), New Hampshire (GSE), and Rhode Island (NE), the incentives are almost always positive since they are accompanied by guarantees on program cost recovery, and by pre-existing explicit or implicit decoupling mechanisms that automatically remove the disincentives associated with reduced sales. In California, however, sub-par program performance, measured by program participation relative to a target value, can lead to earnings penalties.

Shared savings are unique from other utility incentives for DSM in that they make the link between the net resource value of demand-side activities and utility earnings explicit. In this regard, shared savings reward utility performance in acquiring cost-effective demand-side resources, rather than spending ratepayer dollars.

A potential disadvantage of basing utility incentives on net resource value is the need to measure this value, in particular, the load reductions resulting specifically from utility demand-side activities. For each of the utility shared-savings incentives examined, estimates of load reductions on a per-measure basis are being used in conjunction with actual program participation levels. In effect demand-side *measure* performance risks have been largely transferred to the ratepayer, while demand-side *program* participation risk remains with the utility. At the same time, significant utility resources are being devoted to measuring and evaluating programs to provide better estimates for future demand-side measure performance. A consequence of agreements to use measure performance estimates, as well as estimates of future avoided costs, is that net resource benefits are largely agreed on in advance and can be quickly recovered by the utilities.

As a result of these agreements, the shared-savings incentives for PG&E, SDG&E, GSE, and NE are very clear and understandable: if the utility can achieve pre-specified performance thresholds, then well-defined incentives will be earned. With the exception of knowing whether it will meet its program performance targets (specified as energy savings or program participation levels), the utility can predict exactly how much it will earn. Accelerated recovery of the incentives also simplifies administration by commissions and the utilities because incentive recovery is completed within a few years' time.

On the other hand, California's shared-savings incentives feature detailed program design elements that tend to complicate their administration. To address cream-skimming and to ensure utility participation in a variety of demand-side markets, California shared-savings incentives include program-by-program performance (i.e., participation) thresholds, below which penalties apply.

The shared-savings incentive for PG&E is based solely on utility costs, not utility and customer costs. While this tends to increase the net resource benefit for which PG&E is eligible to earn a percentage, it also provides a strong signal for the utility to minimize its own costs (reducing rate impacts) although not necessarily the customer's cost in acquiring demand-side resources. For example, partly as a result of this decision, PG&E rebates typically pay only a fraction of the incremental costs of an energy efficiency measure. In contrast, the NEES subsidiaries whose shared savings are based on total costs typically pay almost 100% of the incremental cost of energy efficiency measures.

The DSM incentives available to GSE, NE, PG&E, and SDG&E also address broader policy considerations for demand-side resources. GSE's and NE's incentives include a "maximizing" incentive that provides additional incentives for demand-side measures with smaller net resource benefits, such as certain residential programs. PG&E's and SDG&E's incentives address these concerns through program-specific performance thresholds and penalties. They also distinguish between classes of demand-side activities and provide separate, non-shared-savings incentives for some of them.

Finally, the results from the utilities' 1990 DSM activities confirm the profitability of the incentives. As a percent of total DSM program expenditures, the incentives are providing measurable returns (PG&E, 11%; SDG&E, 60%; NEES, 12%). At the same time, incentives to utilities for their DSM activities also measurably increase society's cost of acquiring DSM.

Acknowledgments

The work described in this paper was funded by the Assistant Secretary for Conservation and Renewable Energy, Office of Utility Technologies of the U.S. Department of Energy under Contract No. DE-AC03-76SF00098. The authors would also like to thank Mark Hutchinson, New England Electric System; Scott Logan and Ali Miremadi, Division of Ratepayer Advocates, California Public Utilities Commission; Steve Wiel, Nevada Public Service Commission; Chuck Goldman and Ed Kahn, Lawrence Berkeley Laboratory; and, finally, the editors of this book for their assistance and helpful comments.

References

Blumstein, C., B. Krieg, C. York, and L. Schipper. 1980. "Overcoming Social and Institutional Barriers to Energy Conservation." *Energy* 5. Pp. 335–371.

Caldwell, C., and R. Cavanagh. 1989. *The Decline of Conservation at California Utilities: Causes, Costs, and Remedies*. San Francisco, Calif.: Natural Resources Defense Council.

California Public Utilities Commission and California Energy Commission. 1987. *Standard Practice Manual, Economic Analysis of Demand-Side Management Programs*. CEC Report P400-87-006. Sacramento, Calif.

Chamberlin, J. and P. Hanser. 1991. "Current Designs of Regulatory Techniques to Encourage DSM." In *National Conference on Integrated Resource Planning*. Santa Fe, N. Mex. April 8–10. Washington, D.C.: National Association of Regulatory Utility Commissioners. Pp. 316–326.

Energy Information Administration. 1991. *Electric Sales and Revenues 1989*. DOE/EIA-0540 (89). Washington, D.C.: U.S. Department of Energy.

Gallagher, J. 1991. "DSM Incentives in New York State: A Critique of Initial Utility Methods." In *DSM Building on Experience. 5th National DSM Conference*. Boston, Mass. July 30–August 1. Palo Alto, Calif.: Electric Power Research Institute.

Hutchinson, Mark. 1991. Personal communication with Joe Eto. Westborough, Mass.: New England Power Service Company System (NEPSCO). November 6.

Krause, F. and J. Eto. 1988. *Least-Cost Planning, A Handbook for Public Utility Commissioners Vol. 2, The Demand Side: Conceptual and Methodological Issues*. Washington D.C.: National Association of Regulatory Utility Commissioners.

Moskovitz, D. 1988. *Profits and Progress Through Least-Cost Planning*. Washington. D.C.: National Association of Regulatory Utility Commissioners.

Ottinger, R., D. Wooley, N. Robinson, D. Hodas, and S. Babb. 1990. *Environmental Costs of Electricity*. Pace University Center for Environmental Legal Studies. New York, N.Y.: Oceana Press.

Pacific Gas and Electric Company (PG&E). 1991. *Annual Summary Report on Demand-Side Management Programs in 1990 and 1991*. San Francisco, Calif.

Rosenblum, B. and J. Eto. 1986. *Utility Benefits from Targeting Demand-Side Management Programs at Specific Distribution Areas*. EM-4771. Palo Alto, Calif.: Electric Power Research Institute.

San Diego Gas and Electric Company (SDG&E). 1991. *Annual Summary of DSM Activities*. San Diego, Calif.

Sant, R. 1980. "Coming Markets for Energy Services," *Harvard Business Review*. May–June. Pp. 1–8.

Schultz, D., and J. Eto. 1990. "Carrots and Sticks: Designing Shared Savings Incentive Programs for Utility Energy Efficiency." *The Electricity Journal* 3(10). December. Pp. 32–46.

Wellinghoff, J. 1988. "The Forgotten Factor in Least-Cost Utility Planning: Cost Recovery." *Public Utilities Fortnightly*. March 31. Pp. 9–16.

Wiel, S. 1990. "The Urgent Need for Verifying DSM Achievements." *ACEEE 1990 Summer Study on Energy Efficiency in Buildings. Vol. 6, Program Evaluation*. Washington, D.C.: American Council for an Energy-Efficient Economy. Pp. 215–223.

Chapter 7

Revenue Decoupling Plus Incentives Mechanism

L. Mario DiValentino, Terry L. Dittrich, James E. Cuccaro, and
Alan M. Freedman

Introduction

It is becoming increasingly apparent that meeting customers' needs
through supply-side options must be balanced by cost-effective
demand-side options. It is equally important that the use of demand-
side options must balance the needs of shareholders with those of
customers.

Recognizing this shift in the regulatory and social climate, the
New York Public Service Commission (Commission), in Opinion No.
88-20, issued July 26, 1988, investigated methods and ratemaking
procedures designed to meet the future energy needs of customers in
the most efficient, least costly manner. Acknowledging that a least-cost
planning process must consider alternatives to the traditional practice
of constructing new generating facilities in response to load growth,
the Commission re-emphasized its commitment to demand-side management (DSM) while recognizing existing impediments to successful
DSM program implementation.

In its Opinion, the Commission recognized that:

- Energy conservation and DSM are principal means of satisfying
New York's energy requirements; and

- Existing ratemaking practices penalize—rather than reward—cost-
effective DSM measures, because the lost sales and profitability from
successful conservation programs pits a utility's shareholders' interests against its customers' interests.

The Commission, concluding that this conflict should be resolved
through revised ratemaking practices, ordered New York utilities to

propose ratemaking innovations that would align the interests of utility shareholders and customers. The Commission's goal was to provide customers with the benefits of least-cost planning and DSM using a mechanism that would also be beneficial to utility shareholders.

Orange and Rockland Utilities, Inc. (O&R) shared the Commission's concerns regarding the inadequacy of existing ratemaking practices to provide the appropriate incentives to promote least-cost planning. The company was experiencing strong load growth coupled with shrinking capacity reserves and was interested in acquiring capacity at the lowest available cost. Adding cost-effective DSM to the more traditional menu of capacity purchases and construction offered an opportunity to meet service obligations to customers with limited risk to shareholders. As a result, O&R began to investigate alternative ratemaking models that would align the interests of shareholders and customers in achieving cost-effective energy conservation while providing incentives to achieve DSM and other important regulatory objectives.

Alternative Ratemaking Models

The Commission noted that ratemaking procedures adopted in California, Maine, Washington and Wisconsin might be considered useful examples for the ratemaking approach under consideration in New York. O&R reviewed the procedures followed in these states and others. While DSM was promoted in various ways, none of the approaches appeared to balance cost-effective DSM and supply-side options. Other notable observations included:

- DSM expenditures were often recovered through a surcharge or reflected in rates through deferred accounting mechanisms;

- In some cases, utilities were allowed to recover revenues from the "lost sales" that resulted from DSM, removing a powerful disincentive for the successful introduction of energy conservation measures; and

- Positive incentives were sometimes offered to utilities for successfully implementing the DSM option.

O & R was most interested in the last point regarding incentive regulation. This concept not only promoted cost-effective DSM and energy conservation, it also provided an opportunity for the utility to deploy resources effectively and efficiently to meet other meaningful regulatory objectives.

While the opportunity to earn a return for shareholders is a powerful stimulus for utilities to pursue a particular course, the type of incentives in the ratemaking models of the other states seemed either too weak or too misdirected. For example, several states provided for including DSM expenditures in the rate base and offered a premium return on these investments. This sort of incentive—or bonus—is tied directly to the level of expenditures incurred, rather than to the successful implementation of the programs. O&R believed that effective incentive regulation should reward the achievement of results, not the level of spending.

The plan receiving the most attention from Commission staff and other interested parties was the Electric Revenue Adjustment Mechanism (ERAM) used in California (see Chapter 3). Since ERAM decouples a utility's profitability from its sales volume, it has created an environment where utilities are not penalized for implementing DSM measures. Under ERAM, energy sales revenues are established in a base rate case and actual sales revenues are then reconciled to this level. All differences between the sales rate allowance and actual sales are deferred for subsequent refund or surcharge to customers. This decoupling procedure is based on the perception that there is a fundamental inconsistency between a utility's corporate goal of maximizing earnings by maximizing sales and the state's energy policy goal of achieving energy end-use efficiency.

O&R had serious reservations about ERAM's practical application in New York. While ERAM removes the financial disincentive to promote energy conservation, its impact on promoting DSM is marginal at best and counterproductive at worst.

The mechanism does not differentiate among the source of the sales variance (including weather-related sales), economic conditions, growth in number of customers served, and energy conservation. The use of such a broad-based reconciliation mechanism, which insulates the utility from the financial impact of actions actually taken by its customers, could potentially create a sense of indifference to customer energy use by both the utility and its regulator. It seemed clear that ERAM alone would do little to encourage effective DSM measures. Indeed, if one assumes that utilities are interested solely in enhancing earnings, it would be illogical to assume that they would make significant expenditures for DSM programs when they are completely insulated from their financial impact.

A thorough examination of the complete California ratemaking system revealed that rather than promoting DSM, ERAM was a component of a framework designed to provide rate stability. The other elements of the California ratemaking system are:

- Energy Cost Adjustment Clause (ECAC), a recovery surcharge for fuel and purchased power costs;

- Annual attrition adjustments for non-fuel operating and maintenance expenses, financing costs and rate base; and

- Mandated three-year rate cases.

It is this last element—the mandated three-year rate case cycle—that is the most critical in the California regulatory system. The difficulty of accurately forecasting energy sales, operation and maintenance (O&M) expenses, finance costs, and rate base for a three-year period led the utilities and regulators to devise annual rate adjustment procedures. These provide rate and earnings stability by reducing the volatility caused by variances between actual and forecasted results. O&R believed that despite its potential to make energy conservation attractive, ERAM is a single element in a complex ratemaking framework primarily designed to implement a three-year rate case cycle. This conclusion did not rule out ERAM's possible use in New York as part of a least-cost planning approach, but it was clear that ERAM alone would not stimulate New York utilities to change direction in terms of DSM.

Development of the Revenue Decoupling Mechanism (RDM)

O&R concluded that none of the ratemaking models used in other states could effectively be used in New York. Therefore, the company decided to propose a new plan—one grounded in traditional New York ratemaking—but incorporating several concepts used elsewhere to reform the existing system in order to stimulate least-cost planning and DSM. O&R initially identified five goals, deciding that any new ratemaking system must:

- Facilitate, not hinder, least-cost planning;

- Promote rate stability, providing the opportunity for using resources based on long-term objectives rather than on short-term financial considerations;

- Maintain the regulatory compact, whereby utilities are provided the opportunity to earn a fair rate of return on invested capital in return for fulfilling their obligation to serve the energy needs of its customers;

- Ensure a simple and understandable regulatory process; and

- Maintain a rate structure to facilitate the achievement of economic and social goals, e.g., avoiding rate design changes that might adversely affect the economic and environmental health of communities.

Despite serious reservations concerning the ERAM procedure, O&R decided to incorporate an ERAM-like revenue decoupling mechanism (RDM) into its proposal. The company recognized that successful implementation of a least-cost plan depended on eliminating the existing disincentives in the ratemaking system. Having removed the reward for maximizing energy sales, O&R had to devise other strategies to maintain profitability. The following elements of the RDM proposal attempt to provide the means of achieving the important corporate goal of profitability:

- Revenue reconciliation is combined with O&M expense and rate base attrition adjustment procedures, establishing productivity gains and cost control as the new keys to profitability; and

- Meaningful performance-based incentives are established to focus utility resources on achieving Commission-approved least-cost planning and customer service objectives.

Thus, the complete concept of RDM plus incentives (RDM Plus) is to use an ERAM-type procedure to eliminate undesirable disincentives in the current ratemaking system while establishing a new set of incentives to redirect utility resources toward achieving the appropriate regulatory objectives. As such, RDM Plus is a multi-year rate adjustment process. As in a traditional rate case, a projected one-year revenue requirement is established. At the conclusion of that year, O&R adjusts its base rates to reflect changes in revenues, expenses, rate base, and the cost of capital. The following sections summarize major elements of this new approach to incentive ratemaking.

Treatment of Revenues

Net margin (revenues less fuel and revenue taxes) on actual sales is reconciled to the level used to establish base rates. Differences are deferred for subsequent refund or surcharge to ratepayers. This procedure defers all differences, regardless of the source of deviation.

Notwithstanding the desirability of DSM, any ratemaking formula must recognize that economic health and growth of the franchise area is a fundamental goal of utility regulation. Consequently, it is imperative that ratemaking not contain disincentives for utilities to meet the demand for service from new customers. DSM should not be promoted at the expense of the economic welfare of the community. To ensure that RDM Plus does not create a perverse incentive, revenue reconciliation was modified to exclude 50% of the customer service charges recovered from new customers added during a given period, thus allowing the company to recover the incremental costs associated with the additional customers.

Cost Attrition Adjustments

The establishment of RDM Plus is only a first step in revising the rate-making system to support least-cost planning while providing the opportunity to achieve the other important regulatory goal of maintaining rate stability. In New York, multi-staged filings were well established, whereby base rates are set and adjusted periodically to reflect changes in specific costs and to provide for their recovery. The increases are allowed on a pre-determined timetable after Commission review. In addition to avoiding expensive, time-consuming base rate filings, this procedure encouraged utilities to control costs and seek out productivity gains in order to operate within the modest revenue increases resulting from the staged filings.

Because the revenue deferral procedure—which decouples sales and earnings—eliminates the contribution that sales growth makes to covering cost increases, RDM Plus provides a mechanism similar to the existing staged filings to recover the increased cost of providing service. Absent timely cost recognition, the revenue adjustments resulting from RDM Plus will inevitably produce a continuous cycle of annual rate proceedings, diverting the Commission and the company from using their resources to improve service cost effectively.

Therefore, to preserve rate stability, RDM Plus expands the cost-control incentives that exist in current New York ratemaking by incorporating a provision for annual rate adjustment to recover increases in the cost of providing service. RDM Plus provides for three basic cost attrition adjustments, which are each discussed below:

• O&M expenses;

• Rate base investment; and

• Cost of senior capital.

O&M Expenses. From the utility's standpoint, one key to profitability under RDM Plus is to control costs of providing service at or below the levels provided by the attrition adjustments. The attrition procedures for O&M expenses are designed to encourage cost control and stimulate productivity.

The specific mechanism used reflects the various types of utility costs. Because no single method is appropriate for all types of expenses, three basic methodologies are used:

- Inflation adjustments are used for categories of costs when the number of units (volume) is predictable and established in a base rate case, but the unit price is subject to inflationary trends. The procedure is to annually adjust the rate allowance for the projected inflation rate. The inflation rate is not reconciled or adjusted for actual results. The company is thereby forced to control its actual costs within the expected general inflation rate. The majority of a utility's expenses are in this category.

- Volumetric reconciliations are used for costs when price is controllable but volume is not, such as purchased power costs and power plant overhauls. An annual target price and cost level is established and a subsequent adjustment (up or down) is made based on actual volumes purchased.

- Full reconciliations are used for costs such as wage rates, property taxes and medical, property and liability insurance. Annual adjustments are made to reconcile the rate allowances for these costs to actual expenditures, and are subject to Commission review for reasonableness.

All these attrition mechanisms have been applied in New York at various times in traditional multi-stage rate proceedings. RDM Plus is unique in that it combines these procedures into a single comprehensive system.

Rate Base Investment. Changes in the amount of net utility plant investment were also considered an element of cost attrition under the revenue/earnings decoupling regime. The procedure included in RDM Plus is straightforward and simple to administer. The revenue requirement allowance is updated annually to reflect forecast additions to net utility plant investment and related increases in depreciation expense. The forecast additions are prorated on the basis of the expected in-service date of the facilities. No reconciliation to actual capital costs and completion dates is performed to create an incentive for the company to meet its costs and schedule objectives. The forecast additions are subject to annual Commission review and approval.

Cost of Senior Capital. In addition to operating expenses and rate base investment, the cost of providing service is affected by changes in the utility's capital structure and cost of capital. Changes in the company's capital structure and cost of debt and preferred stock are updated annually. These changes are currently reviewed by the Commission outside of base rate cases through financing petitions and other required filings. Again, RDM Plus takes advantage of existing regulatory procedures by incorporating them into a single system. The base common equity return is established in the context of a base rate case and is unchanged by RDM Plus. However, the total equity return reflects the company's ability to achieve predetermined incentive goals.

Performance Incentives

In O&R's view, meaningful incentives to encourage specific utility actions are the essence of RDM Plus. The absence of such incentives would merely make the utility indifferent from a financial perspective to least-cost planning and DSM. Because utilities react to the economic environment in which they operate, complete regulatory reform can only be achieved by replacing embedded incentives for sales growth and new plant construction with a new set of financial incentives that make least-cost planning and DSM the most profitable course of action. For incentive regulation to achieve the desired results, the incentives must have three basic characteristics:

- The incentive must be meaningful and must have significant value to the utility and its customers if it is to be useful in encouraging specific actions;

- The incentive must be designed to reward the achievement of specific results, rather than levels of expenditures or number of DSM programs implemented; and

- Performance in achieving results must be measurable, and specific measurement methods must be predetermined and agreed upon before the incentive is implemented.

Offering incentives and penalties related to specific utility action is not a new concept in New York. It has been applied for many years to fuel cost recoveries, sales to other utilities and sales to gas-interruptible customers. The difference with RDM Plus is that the incentives have broad-based objectives, rather than limited quantitative goals, for a particular type of transaction. The objectives underlying RDM Plus are:

- Implementation of a least-cost plan with attention to achieving specific DSM goals; and

- Maintenance of the quality of service provided to ratepayers.

Each incentive is an integral part of RDM Plus and each meets the fundamental objectives of the ratemaking mechanism. These incentives are now discussed in detail.

DSM Incentive

The DSM performance incentive ranges from a reward of up to 90 basis points (approximately $2.8 million at the 1991 revenue requirement level) in the authorized return on common equity to a penalty of 20 basis points (worth approximately $622,000). The amount of the incentive/penalty awarded is scaled to levels at which the company achieves its energy conservation and net resource savings targets.

Net resource savings are defined as the total benefits of avoided energy use, including reduced environmental impacts, less DSM program costs. The incentive formula is based on the amount of energy saved and the cost of achieving the savings. Thus, the utility has a strong stimulus for developing and implementing DSM programs that maximize energy savings cost effectively.

DSM Goal Setting

The goals for the DSM performance incentive were based on the 1989 New York State Energy Plan, in which the state established an energy efficiency goal of reducing electricity consumption by 10% compared to expected load growth by 2000. This goal of a 10% efficiency gain achieved over a ten-year period is equivalent to an incremental energy use reduction of 1% per year.

The focus at O&R on least-cost planning signaled the beginning of the company's attention to energy, as compared to power, as the primary objective in its DSM efforts. Previously, O&R focused principally on peak load reduction. As a summer peaking utility with an overall system load factor of approximately 50%, O&R was (and still is) committed to promoting DSM programs to eliminate incremental capacity purchases. The company's peak shaving and load-shifting objectives were the foundation of its old DSM plan. However, the new emphasis on energy efficiency to avoid baseload plant construction and to promote environmental quality found a place in the development of O&R's DSM performance goals.

To incorporate programs capable of achieving the targeted level of energy savings, O&R introduced three new DSM programs addressing efficiency improvements in small commercial and industrial facility lighting, residential lighting, and residential water heating technologies. These measures were selected because they provided significant energy savings to customer classes that previously did not participate

Table 7-1. Orange and Rockland Utilities, Inc.—New York Peak Load Reduction and Energy Efficiency Goals, 1990 and 1991 DSM Plans

| Years | Peak Load Reduction (MW) | | | |
	1990 Plan	1991 Plan	Change	% Change
1991	29	42	+13	45%
1992	35	53	+18	51%
1993	41	63	+22	54%

| Years | Energy Efficiency (MWH) | | | |
	1990 Plan	1991 Plan	Change	% Change
1991	16,257	32,000	+15,743	97%
1992	22,042	67,000	+44,958	204%
1993	27,843	104,000	+76,157	274%

in DSM programs. A comparison of O&R's DSM goals derived from its 1990 and 1991 DSM plans (presented in Table 7-1) illustrates the increased importance of energy efficiency savings in the company's resource acquisition strategy.

Structuring the DSM Incentive

O&R recognized at the outset that setting DSM goals would not be enough to create a truly complete incentive formula. An incentive based solely on the level of energy savings obtained suggests that spending any amount to capture that energy savings would be acceptable. However, the degree to which a utility achieves DSM savings cost effectively must also be considered. The cost-effectiveness feature was captured in the incentive formula through the creation of a two-dimensional matrix, where both DSM savings and cost-effectiveness are interrelated. This DSM performance matrix is illustrated in Table 7-2.

Each axis of the matrix represents the percentage of the two goals achieved for that year. The cells of the matrix present the number of basis points earned. Therefore, achieving 100% of the DSM goal does not necessarily mean that the maximum possible incentive could be earned. One hundred percent of the net resource savings goal would also have to be achieved to obtain that outcome.

Cost effectiveness for O&R is determined through the net benefits or net resource savings of a DSM program. Aggregate program costs are subtracted from aggregate program benefits to get net benefits or

Table 7-2. DSM Performance Matrix: Number of Basis Points Earned

		Percent of Energy Efficiency Goal Achieved											
		110	100	90	80	70	60	50	40	30	20	10	0
Percent of Net Resource Savings Goal Achieved	110	90	90	89	79	69	59	50	0	−5	−10	−15	−20
	100	90	90	81	72	63	54	45	0	−5	−10	−15	−20
	90	89	81	73	65	57	49	41	0	−5	−10	−15	−20
	80	79	72	65	58	50	43	36	0	−5	−10	−15	−20
	70	69	63	57	50	44	38	32	0	−5	−10	−15	−20
	60	59	54	49	43	38	32	27	0	−5	−10	−15	−20
	50	50	45	41	36	32	27	23	0	−5	−10	−15	−20
	40	40	36	32	29	25	22	18	0	−5	−10	−15	−20
	30	30	27	24	22	19	16	14	0	−5	−10	−15	−20
	20	20	18	16	14	13	11	9	0	−5	−10	−15	−20
	10	10	9	8	7	6	5	4	0	−5	−10	−15	−20
	0	0	0	0	0	0	0	0	0	−5	−10	−15	−20

net resource savings. The calculation of projected net resource savings in O&R's approved DSM plan provides the net resource savings targets expressed in the DSM performance matrix.

At this point, one additional step is necessary. O&R could potentially be in a situation where more DSM opportunities might be available, but because the DSM budget has been spent, the company would choose to forego the incremental energy savings to meet the cost-effectiveness target. For example, having spent $100 to secure 100 units of energy efficiency, there would be no incentive to spend an additional $50 to secure another 100 units, should that opportunity arise.

To avoid potential lost opportunities, O&R will determine the percentage of the DSM goal achieved by each program and adjust accordingly the projected costs of each program and its net resource savings target. These adjustments will allow a comparison of actual net resource savings to the amount which *should* have been obtained at a given level of savings. Therefore, O&R will not be penalized for taking advantage of the opportunity of acquiring additional energy savings at approximately the same "unit cost."

Applying the incentive mechanism at the end of each year during the three-year cycle means that DSM program impacts and costs must be determined to calculate the earned incentive. Incentives are then rewarded the year after goals are met.

Determining DSM Program Impacts

Determining the success of DSM programs is extremely difficult. The task of evaluating the impact of a DSM program is one of defining the difference in the demand and energy consumption of a participating customer as a result of the utility program as compared to what the customer would have done in the absence of the program. Clearly, measuring DSM savings is very different from measuring supply-side alternatives. Meters cannot always capture what the participant's behavior would have been without utility incentives. Since impact evaluation issues are so important and their methodologies are still under development, it would probably be counterproductive to require absolute perfection in the process of determining DSM savings.

To avoid having uncertainty paralyze the process, O&R proposed and received approval to implement a savings determination strategy using predetermined measurement criteria. These criteria are based on algorithms that precisely state the methodology to be used to quantify program impacts. These algorithms are derived from the most current information on the parameters that influence the impact of DSM programs on electricity usage.

The O&R savings determination strategy also allows for periodic updates to the measurement criteria as more information, knowledge and experience are gained through program implementation and evaluation. The measurement criteria are updated annually, if necessary, and used the following year. As a result, O&R staff is aware of the measurement criteria applied to each DSM program and can allocate the resources accordingly to implement programs efficiently and cost effectively. In addition, the savings determination strategy is not stagnant but changes over time to accommodate the advances in DSM measurement data and techniques.

Implications for DSM

The implementation of RDM Plus has successfully removed all of the disincentives usually associated with DSM. RDM Plus provides for the timely recovery of program costs, recovery of the lost revenues associated with DSM success, and a stimulus for utility management to aggressively pursue DSM savings for its customers. The substantial increase in O&R's commitment to pursue DSM to meet customers' energy needs is a clear reflection of the power of performance incentives.

The strength of the DSM performance incentive is related to its simplicity and pragmatism. The incentive is straightforward and well defined, and its components are easily applied and measurable. All these attributes enhance the mechanism's ability to perform. Ultimately, however, the mechanism will be evaluated based on the benefits delivered to the electric system and the customers served by that system.

While O&R's DSM performance incentive mechanism is achieving its purpose, some modifications are possible. First, peak reduction could be included as a goal in addition to energy efficiency. Secondly, the incentive could include a means to reward the company for exceeding established energy efficiency goals. Finally, while the mechanism does have performance tests to determine program results, it could be modified to encourage the pursuit of long-term energy efficiency measures.

Customer Service Incentive

Another component of O&R's incentive ratemaking approach is the inclusion of a customer service incentive. This incentive is calculated on the basis of the company's performance in six categories and results are measured monthly using a point system. With the achievement of the maximum number of points, the company has the opportunity to earn 16 additional basis points on common equity return (worth approximately $498,000 at the 1991 revenue requirement level). Conversely, a corresponding penalty of a 16-basis-point reduction in common equity return would accompany a score of zero. The annual point allocation is:

- Customer inquiry response (12 points);

- New construction service installations (12 points);

- Emergency response time and meter installations (12 points);

- Non-emergency repair time (12 points);

- Billing accuracy (12 points); and

- Number of Commission complaints (36 points).

Some have argued that the customer service incentive is unnecessary and is like rewarding the utility for doing what it should do routinely. However, with the adoption of the RDM Plus mechanism, one of the company's primary priorities will be to continue maximizing productivity and reducing costs. The customer service incentive encourages the company to forgo short-run savings at the expense of long-range benefits, thereby making it consistent with least-cost planning.

Conclusion

RDM Plus was created through a cooperative effort involving O&R, the Commission and other interested parties. The common goal was to design a comprehensive incentive ratemaking mechanism that focuses regulatory and utility resources on providing energy at the lowest possible cost and on stimulating meaningful DSM efforts. To date, the company has achieved its DSM goals using its marketing department's delivery system and using energy service companies acquired through competitive bidding. Further, the company has earned the entire return on equity incentive for the first 12 months implementation of RDM Plus.

The revenue decoupling mechanism does not represent a radical departure from the current ratemaking in New York; many of its components have previously been used by the Commission. RDM Plus does represent an effort to combine the most desirable features of traditional ratemaking with a regulatory incentive system that produces a single comprehensive mechanism to provide tangible benefits to utilities and their customers.

The result of this effort is an innovative ratemaking mechanism with benefits and applicability that extend well beyond its original scope. In addition to revising ratemaking procedures to be more consistent with DSM and least-cost planning, the implementation of RDM Plus presents opportunities to:

• Provide rate stability;

• Key profitability to increased productivity and cost controls;

• Discourage ratemaking gamesmanship;

• Continue to hold utilities at risk for items over which they do exercise direct control; and

• Establish incentive regulation as the principal means of ensuring achievement of important regulatory objectives.

O&R is convinced that in time these objectives will be fully met and will demonstrate that RDM Plus's value extends well beyond providing an incentive for DSM. As results are measured and as procedures are refined, the revenue-decoupling-plus-incentives mechanism will be an increasingly useful ratemaking tool to align the interests of customers and shareholders.

Chapter 8

Bill Indexing

David Moskovitz and Richard Rosen

Introduction

This chapter describes some of the policies and the practical implications of using customer bills as a simple yardstick to measure a utility's overall success in implementing a least-cost plan generally and, more specifically, using them to structure regulatory incentives to stimulate least-cost planning. Bill indexes and their variations, their strengths and weaknesses, and lessons learned from regulatory experience to date are discussed.

What Is a Bill Index?

A bill index is a benchmark based on utility customer bills to which the customer bills of a "target utility" can be compared. The appeal of bill indexing is that comparing the changes in the average bills of its customers to changes in the bill index, or benchmark, may be a relatively simple means by which regulators can gauge a utility's least-cost planning performance. This approach is particularly useful when taking a long-term perspective. *Bill indexing approaches are not intended to measure the actual impact of specific DSM programs or supply-side improvements. They are intended to provide a relative yardstick to encourage and distinguish superior performance in least-cost planning.*[1] Bill indexing is not an incentive mechanism, per se, but can be an ingredient in several possible mechanisms, including shared savings, rate-of-return adjustments, or a bounty. Some examples of how this can be done are discussed later in this chapter.

A bill index may be either "external" or "internal." Within these

---

[1] Lowering bills and revenue requirements, as distinguished from lowering prices.

two general categories there are also other possible variations, each of which has properties that are more or less suitable depending on jurisdictions and circumstances.

External Bill Index

With an external bill index the benchmark is constructed using data that are entirely *exogenous* to the target utility. There are at least two ways to develop an external bill index:

Option 1: Utility Performance Relative to Other Utilities. This methodology was first suggested in a paper presented at the National Association of Regulatory Utility Commissioners' (NARUC) First Least-Cost Planning Conference (Moskovitz 1988). With this approach, actual average customer bills of a target utility are computed and compared to the average customer bills for a selected group of comparable utilities.[2] Least-cost planning performance by the target utility would be evaluated using changes in its average customer bills compared to those of the comparable group of utilities.

It is important to select the benchmark group of utilities so the aggregate, or average, operating environment for the group as a whole is approximately equivalent to that of the target utility. This similarity controls for factors for which the target utility is not held accountable. For example, if in the aggregate the comparable group of utilities had a fuel mix similar to that of the target utility, subsequent changes in fuel prices would produce approximately the same effects on the average customer bills of both groups. Consequently, the target utility would be insulated from changes in average bills due to changing fuel prices.[3]

Likewise, the target utility would be protected from bill changes caused by weather if both the comparable group and target utility had similar weather conditions and similar levels of weather-sensitive loads such as space-heat or air-conditioning. Variations in the level of customer bills caused by weather would not result in a change in the *relative* level of customer bills.

In contrast, bill impacts at the target utility resulting from more aggressive cost-effective DSM, superior supply-side acquisitions, and improved purchasing practices or plant operations (e.g., heat rate or

---

[2] For example, the Niagara Mohawk Power Corporation uses a statistical technique called "simulated annealing" to identify utilities with comparable cost and sales behavior (Lowry et al. 1991).

[3] Differences in other factors besides price, such as average heat rate, will still affect relative index values between utilities.

forced outage rate) will be reflected by a relative improvement in average customer bills compared to the control group.

Several variations of this indexing option have been examined in research conducted in New York State on behalf of Niagara Mohawk Power Corporation (NMPC) and in Maine for Central Maine Power Company (CMP). Results of these analyses are discussed later in this chapter.

Option 2: Customer Bills Compared to Econometric Forecasts of Utility Bills. This method is a technically sophisticated extension of the previous option. Using this approach, an econometric equation is used to predict the average customer bills of a target utility. The target utility's performance in least-cost planning is measured by comparing *actual* average customer bills for a target utility with the average bills *predicted* by the econometric equation.

The econometric equation is developed by examining data for a large group of utilities and identifying the impact that "key cost drivers" (independent variables in the equation) have on the utilities' average bills. The equation is then used to predict the average bill for the target utility.

The construction of the econometric equation is critical. Exclusion or inclusion of specific independent variables ("drivers") will determine whether the utility is or is not held accountable for effect that these variables may have on customer bills.[4] Utilities are only held accountable for changes in the dependent variable (bills) which are the result of factors *excluded* from the econometric equation. Utility least-cost planning performance as well as overall managerial efficiency are presumably examples of such excluded variables.

For example, if regulators do not want to hold utilities accountable for changes in average customer bills resulting from changes in fuel prices, fuel prices would be included as one of the variables in the econometric equation. The econometric equation would then use actual fuel prices as one of the explanatory variables that determine average bills for the target utility. In this way, changes in average bills due to fuel price deviations would not expose the target utility to excessive rewards or penalties.

Internal Bill Index

Under an internal bill index, a benchmark is constructed using information pertaining *solely* to the target utility. No comparisons are made

---

[4] Efforts at Niagara Mohawk to forecast average residential bills incorporated six independent variables: heating and cooling degree days, regional price index, personal income, state income tax rate, and excess demand (Lowry, et al. 1991).

to data or performance of other utilities. There are also at least two ways to construct an internal bill index:

Option 3: Average Customer Bills Compared to Average Bills Predicted by an Approved Least-Cost Plan. Many states require utilities to engage in least-cost planning and to submit periodic utility filings followed by a formal plan approval process. Implicit in the approval of plan is an estimate of future average customer bills that are expected to result from plan implementation.

Under this option, a utility's success in achieving least-cost goals would be measured by comparing actual average customer bills with the value of average bills forecasted in the least-cost planning exercise.

As is the case with the first two options, this approach can be refined to limit the utility's exposure to specific factors. For example, any forecast of average customer bills derived from a least-cost plan will, in part, be based on forecasted fuel prices and normal weather conditions. To the extent that regulators do not intend to hold utilities accountable for deviations between *actual* versus *forecast* average customer bills due to fuel price or weather variability, the forecasted average bills should be recomputed based on actual fuel prices and weather.

Option 4: Average Bills of the Customer Control Group Compared to Average Bills of All Customers. With this approach a control group of a utility's customers is selected. Various criteria ensure that this sample group will provide statistically sound comparisons with the entire population of all ratepayers. These criteria are:

- The group should be representative of the overall population of the utility's customers in all regards, including usage characteristics, customer size, and participation in prior utility energy efficiency programs; and
- The control group should be large enough to allow reliable detection of relatively small differences between the average bills of the control group and those of the overall population of customers.

Given these criteria, average customer bills for the control group at the beginning of the test period should equal bills for the overall customer population. Similarly, because rates charged to both groups are equivalent, equal average bill means that the average kWh use of the two groups is the same. During the test period, participation in DSM programs by customers who are not in the control group will produce a reduction in their bills compared to those in the control group. The kWh savings of the utility's DSM program during the specified period can be calculated as follows:

$$\text{kWh saved} = (\text{AkWh}_{control} - \text{AkWh}_{overall}) * \text{NC}$$

where:

$\text{AkWh}_{control}$ = Average kWh use of customers in the control group

$\text{AkWh}_{overall}$ = Average kWh use of all customers

NC = Total number of customers.

The utility would be rewarded or penalized based on the computed kWh savings compared to a target level of savings for the same period.

This indexing approach is the most limited of the four options described because it only measures the energy savings performance of the DSM components of least-cost planning. Neither DSM cost-effectiveness nor supply-side resource performance is addressed.

Common Elements

Each of the four bill indexing options previously described provides a relative measure of customer bill savings and LCP performance that can be used to design specific incentive plans. The structure of an incentive plan is flexible. Shared savings, rate-of-return adjustments, or bounty plans can be designed using bill indexes as the yardstick of performance. For example, under Option 1, the identified bill savings can be expressed in total dollars, average bill savings times number of customers, and shared among customers and shareholders using a reasonable proportion. Similarly, the share of the utility's savings can be expressed in terms of a rate-of-return adjustment ("x" basis points for each percentage point of bill savings) or a bounty ("y" dollars for each percentage point of bill savings). Regardless of the structure, the level of the incentive available to the utility must be limited by identified reductions in utility revenue requirements.[5]

Is the Use of Customer Bills Compatible with Least-Cost Planning Goals?

The goal of least-cost planning is to meet the energy service demands of customers at the lowest possible cost and with the lowest long-term revenue requirements. Given this objective, the use of average long-

---

[5] The maximum utility incentive is further limited by regulators' judgments regarding reasonable limits on overall financial return. Experience to date suggests that whether an incentive plan is structured as shared savings, bounty, or rate of return, an overall incentive ceiling on the return on equity (perhaps 100 basis points) may be appropriate.

run changes in customer bills may be a reasonable, although imperfect, measure of a utility's least-cost planning performance.

In examining the efficacy of a bill index system we must also ask to what extent minimizing the costs of energy services is consistent with minimizing customer electricity bills. Least-cost planning tries to minimize the cost for a given energy service demand over the long-term. In some cases, the least costly manner of meeting an energy service demand may mean increased, rather than decreased, use of electricity. In addition, least-cost planning evaluation incorporates other "costs" that are not reflected in customer bills, such as customer, as opposed to utility, contributions for installing DSM measures. With these considerations in mind, what are the implications of relying solely on a bill index to measure least-cost planning performance?

The relationship between customer bills and the cost-effectiveness tests used in the least-cost planning process helps provide insight into these questions.[6]

For example, in jurisdictions where the non-participant or rate impact test is the primary test for cost-effectiveness, customer bill yardsticks are unsuitable. To satisfy this test, utility actions must seek to minimize electricity *prices*, not customer bills. If average prices exceed marginal or avoided costs, higher sales will yield lower prices. Assuming no change in the number of customers, however, higher sales produce higher customer bills even when prices are reduced. If the regulatory objective is to minimize prices, using a bill index to reward the utility for lower customer bills will be counterproductive.

On the other hand, a bill indexing mechanism is completely consistent when the utility cost test is applied and the planning goal is to minimize utility revenue requirements. Because a utility's average customer bill equals its revenue requirement divided by the number of customers, the objective of minimizing revenue requirements is equivalent to minimizing customer bills.

In recent years neither the non-participant nor the utility cost test has been widely used as the primary test of cost-effectiveness. Least-cost planning jurisdictions typically employ the broader total resource cost test (TRC); the regulatory objective in this setting is to minimize the totality of *all* resource costs, including those borne by the customer as well as those paid by the utility. Thus, the question in this context becomes whether customer bills (or some other measure) are a reasonable yardstick of total resource costs. Stated another way, what are the implications of using utility costs as a proxy for total resource costs?

---

[6] A useful description of these cost-effectiveness tests can be found in California Public Utilities Commission and California Energy Commission 1987.

Total Resource Cost

Since total resource costs (TRC) include DSM costs incurred directly by customers, by definition these customer costs are not included in customer electric bills. For example, if a utility rebates half the cost of a DSM measure to a customer, the rebate expenditure alone is included as a cost that is ultimately reflected in bills. If, on the other hand, the utility routinely paid the entire cost of the DSM measure, the utility's incremental revenue requirement and the total resource cost would be almost identical. In this case, the utility's success in meeting the energy service needs of its customers at the lowest total resource cost will also minimize its revenue requirements and its average customer bill. However, this last scenario is the exception. In most utility-sponsored DSM programs, customers contribute a substantial proportion of the costs of installed efficiency measures.

The exclusion of customer expenditures for energy services from average customer bills makes bills a somewhat imperfect yardstick of utility performance. For example, with a customer bill yardstick, the utilities would have an incentive to pursue DSM programs that failed the TRC test but whose customers paid enough of the cost of the measure to reduce electric bills. Of course, compared to the incentives inherent in conventional regulation that must be overcome through regulatory oversight, the risk of utilities pursuing significant amounts of non-cost-effective DSM with high customer contributions is not serious.

This imperfect alignment between customer bills and total resource cost may, surprisingly, also be a positive feature of a bill indexing approach if other steps are taken to assure that only cost-effective DSM programs are pursued. The greatest overall reduction of average customer bills is achieved when utilities pursue all cost-effective DSM *and* obtain the largest possible contribution from participating customers. These dual and competing goals to maximize the penetration of cost-effective DSM and get participating customers to pay as much as possible, parallel the balance that regulators make when relying on the total resource cost test as the primary least-cost planning test, while remaining mindful of rate impacts resulting from DSM and other least-cost planning activities.[7] Thus, a customer bill yardstick, while less than a perfect measure of total resource costs, may reconcile individual ratepayer interests with broader regulatory

---

[7] Competing goals are common. For example, regulators and utilities have historically balanced the desire for high system reliability against minimizing cost.

Table 8-1. Price and Bill Impacts

Assumptions:

Average price   7 cents/kWh

New DSM and supply additions equal 10% of existing annual sales (1 million kWh), reflecting 10% load growth net of DSM due to customer growth.

Full cost of DSM, lost revenues, and supply additions reflected in rates.

Action	Revenue Requirement	Price Impact	Bill Impact
Base case prior to 10% customer growth	$70,000	—	—
Implement DSM at cost of 3 cents/kWh	$73,000	4%	−5%
Add supply at cost of 3 cents/kWh	$73,000	−5%	−5%
Implement DSM at cost of 5 cents/kWh	$75,000	7%	−3%
Add supply at cost of 5 cents/kWh	$75,000	−3%	−3%

Note: This table is based on examples where average prices exceed avoided (supply) costs. The general conclusions do not change when prices are less than avoided cost.

policies and objectives more successfully than a theoretical measure that was more consistent with the total resource test.

Table 8-1, based on hypothetical assumptions, examines the impact of various supply and demand-side actions on average customer bills and average prices.

Table 8-1 illustrates several points. First, revenue requirements and *average bills* (revenue requirements divided by number of customers) are unaffected by whether the chosen resource is demand-side or supply-side. Second, *average prices* are influenced more by whether the new resource is a supply- or demand-side resource. Third, as resource costs decrease, *both* average bills and price impacts decrease.

Two additional attributes of customer bill measures are worth noting. Under a system of bill indexing, customer bills will reflect *actual* instead of *estimated* DSM savings. If DSM measures fail to produce actual savings, the savings will not be reflected in average bills. Customer bills of the target utility are easily measured and verified, and only actual DSM impacts (both actual costs and actual energy savings) and other changes in efficiency are reflected in bills. In a similar fashion, DSM programs that produce greater savings than anticipated are automatically reflected in average customer bills.[8] If utility earnings are linked to changes in average bills, this provides an obvious incentive to implement effective DSM programs.

Customer bills also automatically reflect supply-side efficiency improvements, such as better fuel procurement practices and improved power plant performance as well as administrative cost savings.

Bill Indexing and Decoupling

Chapter 1 examines the importance of decoupling utility profits from sales, and Chapters 3 and 4 describe two decoupling options.

The combination of a reward for lower bills and a penalty for higher bills theoretically means an indexing-based incentive plan can decouple profits from sales. If the incentive received by the utility due to lower bills is larger than the loss of incremental profit associated with the reduced level of sales, the index decouples. Likewise, the plan decouples if the penalty for higher bills caused by a lack of effective DSM programs or the existence of a sales promotion program is greater than the profit earned on the higher sales.

Table 8-2, however, indicates that the financial incentive would have to be very large to overcome the impact of lost revenues. It may, therefore, be difficult to implement a bill-index-based incentive plan with rewards and penalties large enough to remove existing disincentives.

Although bill indexing is theoretically capable of decoupling utility profits from sales, the design and implementation of a bill indexing plan is simplified if bill indexing is combined with a separate decoupling approach. Removing the issue of lost revenues would leave the different bill-indexing approaches with two primary functions: (1) to motivate utility management to focus on customer bills, and (2) to enable a utility's incentive plan to directly promote the pursuit of least-cost planning in general, and DSM in particular.

As a final note on the usefulness of a "stand alone" bill index, research to date suggests that bill-indexing approaches are best suited to *long-term* incentive planning and thus may match the long time-frame inherent in LCP. In contrast, decoupling is a regulatory reform aimed primarily at removing short-term disincentives. Combining the

---

[8] Depending on which of the four options is used, customer bill savings may also be net of the free riders (participating customers who would have invested in DSM measures regardless of utility DSM) and free drivers (customers who invest in DSM measures on their own but would not have done so but for utility programs).

Table 8-2. Decoupling Example		
Assumptions:		
Retail rate		7¢/kWh
Avoided fuel cost		3¢/kWh
"Lost revenue"		4¢/kWh
DSM cost		2¢/kWh
Average annual bill before DSM @10,000 kWh/year		$700
DSM annual savings		2,000 kWh
Bill Indexing	**With Decoupling**	**Without Decoupling**
Average bill after DSM	$680	$600
Lost revenue	$0	$80
Resource savings	$20	$20
Bill savings	$20	$100
Stockholders' total incentive (percentage of bill reduction)*	10%	82%

*This net value is comprised of 10% of resource savings (in this case, $2) plus lost revenue divided by the reduction in the average bill after DSM.

two approaches may provide the most positive and powerful regulatory environment.

Interpreting Results of the Four Bill-Indexing Approaches

Each indexing option compares the average customer bill of the target utility to a different benchmark. Each option, therefore, yields a different measured relative bill reduction, and each computation will have a different meaning.

With the possible exception of Option 4, none of the bill-indexing options is intended to measure actual DSM or supply-side savings. Instead, bill indexing is a comparative yardstick for judging relative performance. The use of bill indexing to evaluate utility performance is similar in this respect to students being competitively graded or the functioning of a competitive market that rewards companies for being more efficient than their competitors.

Table 8-3. General Interpretation of Measured Bill Savings

Option 1	Measures the extent to which the target utility outperforms a group of other utilities, whether or not they pursue LCP.
Option 2	Measures the extent to which the target utility outperforms predictions produced by econometric modeling.
Option 3	Measures the extent to which the target utility outperforms its own least-cost plan.
Option 4	Measures the extent to which utility-sponsored DSM programs reduce average energy consumption.

Table 8-3 summarizes the general interpretation of the numerical measure of bill reductions produced by the four options.

As an example, assume the bill-indexing plan described in Option 1 shows that the average customer bill for the target utility falls by 2% relative to the average bill for the composite group of utilities and that an average bill savings of this size equals a reduction of $100 million per year in the target utility's revenue requirement. Further assume that $60 million of the $100 million savings was created by demand-side efforts and $40 million by supply-side improvements.

In the context of an Option 1 bill-index approach, this result is interpreted to mean that the utility produced $60 million *more* in DSM savings than was achieved by the composite group of utilities used as the benchmark. If the average bill savings for the composite group's DSM efforts were equivalent to $50 million, the target utility would have had to produce $110 million of savings to show a *net* improvement of $60 million.

If the identical utility behavior that produced $100 million savings as measured by an Option 1–type bill index were measured using Option 3, the measured savings could be entirely different. In this case, if $100 million in savings had been specified by the least-cost plan, an Option 3 bill index would measure zero change in customer bills.

If the same utility actions were measured using Option 4, the savings would not show the $40 million of supply-side efficiencies.

Attributes of Bill Index Alternatives

Are They Understandable, Predictable and Simple to Administer?

Understandability. All four of the bill-indexing approaches seem amenable to being understood by utilities, regulators, and the public. Options 1, 3, and 4 are probably the most easily understood.

Option 2 (comparison to econometric forecast) may be the most difficult of the bill-indexing approaches for the public to understand because of the complex multi-termed equations that are used to predict customers' bills.

Predictability. Predictable rewards and penalties will motivate utility managers more than uncertain rewards. For example, if each compact fluorescent bulb installed increases utility earnings by five cents, the plan produces predictable results.

If, on the other hand, the incentive plan rewards the utility for actual energy savings instead of merely for installed measures, the risk of energy savings performance is borne by the utility. Selecting the latter approach means that performance is being weighted more heavily than predictability.

In general, bill-indexing approaches emphasize measured or actual performance and are therefore less predictable than most other incentive strategies.

Predictability is also sacrificed in indexing plans that rely on comparisons to other utilities. For example, utility managers who implement an aggressive DSM program may earn no incentive if utilities in a benchmark group performed as well. Of course, if the utility managers had not implemented a major DSM effort they would have compared poorly to the benchmark and earned a penalty. While it may be impossible to predict whether the overall, or net, incentive associated with particular actions will be a reward or a penalty, the incremental effect of incremental action is predictable.

Administrative Simplicity. Administering external bill-indexing options could be relatively simple once the systems are established. At the outset, however, Options 1 and 2, the comparable utilities and econometric forecast approaches respectively, may be difficult to establish, particularly if opposed by one or more major parties. The level of difficulty of establishing an acceptable yardstick is similar to the difficulty of creating other regulatory mechanisms, such as performance-based fuel and purchased power clauses for electric companies or price cap experiments for telephone companies.

Once comparable utilities or econometric forecasting equations are established, administering the incentive systems should be reasonably simple.

Both internal index approaches would be substantially easier to establish than external approaches, and both internal bill approaches should be easy to administer. Variations of Option 4 are already used by some states and utilities for DSM program measurement.

Cost Control/Non-Participant Impacts/Cream-Skimming

With the partial exception of Option 4, bill indexing provides inherent incentives to control costs and balances concerns of non-participants and not encouraging cream-skimming.

Because DSM costs are reflected in customer bills, a utility will have an incentive to minimize its DSM program cost. And, because any DSM program that costs less than the utility's avoided cost will reduce customer bills, the utility will also have an incentive to pursue cost-effective DSM. Even DSM opportunities that are barely cost-effective and fully paid for by the utility will result in customer bill savings.[9]

Assuming the utility's ability to finance and deliver DSM programs is not constrained, there is no financial reason or incentive, under a bill-index scheme, for a utility to limit its DSM programs to the very lowest cost opportunities, i.e., to skim cream.

Long-Run Versus Short-Run Efficacy

The difficulty in detecting relatively small differences in average bills from *annual*, as opposed to *cumulative* data, suggests that bill indexing may be best suited as a long-run incentive plan. This conclusion is reinforced by the existence, in the short run, of gaming opportunities related to differences in the timing of utility cost recovery.

Table 8-4 shows the impacts on average customer bills of a typical DSM effort (15% of electricity saved in the tenth year) using various cost-recovery schemes.

Table 8-4 shows that the annual changes in average customer bills resulting solely from DSM activities are relatively small (less than 2% per year) even with large DSM investment levels, and that the timing of bill reductions is somewhat influenced by DSM cost-recovery

---

[9]   If the societal cost test for DSM is used that includes externalities, this may not be true unless the externalities are collected like a tax as part of revenue requirements. However, because in most cases customers will directly contribute to the costs of DSM programs, this is unlikely to be a problem.

Table 8-4. Annual Bill Reductions

Year	DSM Costs Expensed	DSM Amortized Using Annual FCF*	DSM Amortized Using Real FCF
1	−1.6%	0.1%	0.2%
5	1.3%	1.0%	1.1%
10	1.7%	0.7%	0.7%

*Annual FCF (fixed charge factors) indicates the use of depreciation practices with a ten-year lifetime. Real FCF, on the other hand, indicates that DSM investment has been amortized using factors levelized in real dollars.

Table 8-5. Cumulative Bill Reductions

Year	DSM Costs Expensed	DSM Amortized Using Annual FCF	DSM Amortized Using Real FCF
1	−1.6%	0.1%	0.2%
5	2.2%	3.2%	3.9%
10	9.8%	7.6%	8.5%

methods.[10] The research thus far suggests that small year-to-year changes in bills will be difficult to detect, particularly when comparing a target utility to other utilities also pursuing LCP. For this reason alone, using a bill index on an annual basis is unlikely to be useful in properly rewarding strong versus weak DSM investment programs.

Table 8-5 is based on the same data as Table 8-4, but shows the *cumulative* change in customer bills. As in Table 8-4, the cumulative change in customer bills becomes much more significant over a period of many years, and is also much less sensitive to the exact cost-recovery mechanism.

This suggests that bill-indexing approaches may be most useful in the context of long-turn regulatory reform plans that will remain in place for an extended period of time. It further suggests that any incentive formula should explicitly be a function of the *cumulative* value of a bill index, not the annual value.

---

[10] Bill reductions resulting from improved supply-side activities in response to regulatory reform are not included in Tables 8-4 or 8-5.

Gaming. The difficulty of detecting small changes in bills from one year to the next, as well as variations in the timing of cost recovery for different resources, means that bill indexing based on annual values creates gaming opportunities on both the demand- and supply-side. These strategies discourage utilities from pursuing genuine least-cost supply options in favor of options with lower costs in early years and higher costs in later years.

Applies to Supply-Side?

Three of the four bill-indexing approaches have the benefit of measuring the performance of both demand- and supply-side activities which is a major benefit. Least-cost supply-side additions or efficiency improvements (improved heat rate, lower forced outage rate, etc.) reduce customer bills to the same extent as equally cost-effective demand-side resources, and therefore should not be ignored in LCP. As shown in Table 8-1, a new supply-side resource acquired at a cost of 80% of the utility's "avoided cost" will affect average bills to the same extent as a demand-side resource acquired at the same cost.

Other Issues

Environmental Costs. By definition, customer bills exclude environmental and other externalities. Thus, incurring costs solely to reduce pollution beyond the legally required minimum environmental standards would not be encouraged by bill indices.

Cost-Effective Use of Electricity. Bill indexing risks discouraging utilities from pursuing opportunities that increase overall customer energy services or increase electricity's share of the energy service market when these objectives are elements of a least-cost plan.[11] Bill indexing discourages these activities because in return for improved or increased service, bills go up.

Summary

Table 8-6 summarizes the attributes of the four bill index options. The column on the left lists the major performance criteria to evaluate alternatives. The notations H, M, and L mean high, medium, and low. N indicates that the criteria are not met.

---

[11] It has been suggested that decoupling also removes the incentive to encourage sales, although in some cases increased sales reduce average prices when prices exceed avoided cost. However, decoupling does *not* prevent utilities from actions that tend to increase electricity consumption. With decoupling, incremental revenues received as a consequence of successful efforts to increase sales flow back to consumers in the form of decreased prices. Decoupling makes utilities neutral towards increasing sales by keeping profits constant.

Table 8-6. Summary of Bill Index Alternatives

Criteria	Option 1	2	3	4
1. Performance-based	H	H	H	H
2. Positive incentive	H	H	H	H
3. Decoupling[a]	M	M	M	M
4. Understandability	H	M	H	H
5. Predictability	M	M	M	M
6. Provision for supply-side resources	Y	Y	Y	N
7. Cost minimization	H	H	H	N
8. Maximizes customer DSM contribution	H	H	H	N
9. Avoidance of cream-skimming[b]	M	M	M	M
10. Inclusion of externalities	N	N	N	N
11. Cost-effective use of electricity[c]	N	N	N	N
12. Simplicity	M	L	M	M

Notes:
[a] It is possible to design a bill-indexing plan that decouples profits from sales, but for reasons discussed earlier it is not recommended.
[b] "Cream skimming" occurs when a utility limits its DSM activities to the easiest and most inexpensive opportunities.
[c] This refers to a possible drawback of bill-indexing options. In some situations there is the risk of deterring utility activities to pursue opportunities that *increase* overall customer energy services or increase electricity's share of the energy service market when this is a valid objective of the least-cost plan.

Research Results to Date

Niagara Mohawk Power Corporation

Consultants working for NMPC in 1991 (Lowry et al. 1991) examined three distinct variants to an external bill index. Two of these three approaches are equivalent to options examined in this chapter.

Econometric Approach. Historical cost and sales data for a large number of investor-owned utilities were taken from FERC reports and used to develop and test alternative econometric equations that could be used to predict NMPC's average residential, commercial, and industrial bills.[12]

---

[12] The researchers examined econometric equations for each customer class, but much of the detailed work was limited to residential bills because of the quality of the FERC Form 1 data, and the problem of consistently defining average commercial and industrial bills.

Potential independent variables were examined for each customer class. For example, in the residential class six substantive variables were ultimately selected in addition to one dummy variable for each year of data, from 1975 to 1988: heating-degree days, cooling-degree days, regional price index, personal income, state income tax rate, and excess demand. Each of the variables selected was statistically significant.

When placed in perspective the statistical performance of the econometric equation for residential customer bills was remarkably good. Overall, the residential bill equation had an R^2 of .53; six independent variables explained 53% of the variation in Niagara Mohawk's average customer bills. The effect of the independent variables on each of the two components of customer bills (price and quantity) was measured. The R^2 for price was .47 and the R^2 for quantity was .59; the customer bill equation was statistically superior to an electricity price equation.

The commercial customer class econometric equation was limited to three substantive variables: input prices, cooling-degree days, and state taxes. The customer bill equation had an R^2 of .51. The R^2 values for commercial price and quantity were .50 and .22 respectively.

The industrial customer-class econometric equation was also limited to three substantive variables: input prices, industrial fraction of sales, and state income taxes. The industrial customer bill equation had an R^2 of .35. The R^2 values for industrial price and quantity were .58 and .22 respectively.

NMPC's study and data showed that using econometric predictions as the foundation of a bill-indexing plan is feasible for the residential customer class and may be feasible for the commercial class. More statistical analysis may be required to develop econometric equations that are better than the three independent-variable equations developed by NMPC. However, the volatility in sales volumes within the industrial customer class and the lack of readily available historical data to improve the predictive power of the equation may make it too difficult to use the econometric bill-indexing approach for the industrial customer class.

Composite Group. The second approach was to identify a composite group of 20 utilities with which to compare Niagara Mohawk's performance. A statistical technique called "simulated annealing" was used to identify this group of utilities with historical cost and sales behavior that most closely matched Niagara Mohawk's.

New York Group. The third approach, a simplified version of the second approach, compared Niagara Mohawk's historic customer bills

Table 8-7. How Rate of Change in Average Residential Bills (% per year) at NMPC Compared with Results from Three Performance Indexes (1983–1988)				
		Index		
Year	NMPC Actual	Comparable Firm	All-Econometric	New York State Utilities
1983–1984	2.90	6.35	4.70	5.63
1984–1985	5.17	3.81	0.75	−0.80
1985–1986	6.75	4.15	1.32	−2.89
1986–1987	3.70	2.59	3.21	0.99
1987–1988	7.34	3.86	3.34	4.08
1983–1988 Average	5.17	4.15	2.66	1.40

to the historic customer bills of five other private New York electric utilities.

Results. The weakness of the econometric approach is that it explains or predicts bill changes *only* in response to changes in the six identified statistically significant independent variables. The strength of the econometric approach is that the independent variables are directly experienced by the target utility.

The area of weakness of the econometric approach is an area of strength for the composite group approach. The customer bills of the composite group respond to *all* business conditions affecting customer bills, not just the few statistically significant independent variables. The main weakness of the approach is that the business conditions that affect the composite group will not affect the target utility.

Table 8-7 shows the performance of the three approaches in matching the historical changes in Niagara Mohawk's average residential bills. For an index to be useful, a good but not perfect fit with historical data is needed; the fit need not be perfect because variations in management efficiency between the target and comparison utilities— one of the factors measured by most indices—will cause year-to-year variations in index values of both target and comparison utilities. As Table 8-7 shows, the larger composite group approach fairly closely matched the actual NMPC results. Table 8-7 also shows, however, that NMPC's actual bill increases exceeded values predicted by the econometric approach, averaging almost 2.5 % per year too high. Bill pre-

Table 8-8. Correlation of per Customer Non-Fuel Costs of Central Maine Power Company and Two Groups of Other Utilities

Regression	R^2
CMP vs. large sample (entire period 1981–1988)	.94
CMP vs. small sample (entire period 1981–1988)	.91
CMP vs. large sample (year-to-year changes)	.09
CMP vs. small sample (year-to-year changes)	.03

dictions based on the New York State utilities produced the largest discrepancies from actual bill changes at NMPC.

Central Maine Power Company

Central Maine Power Company also conducted statistical research that led to a specific proposed incentive plan filed with the Maine Public Utilities Commission in December 1990. CMP's proposal had two parts; one included an external indexing proposal.

The indexing proposal consisted of several components and was based on a statistical analysis of eight years of non-fuel cost data for CMP and two groups of utilities. One group of utilities, called the "small group," consisted of 41 utilities with service and size characteristics similar to CMP. The second group, the "large group," included all investor-owned utilities over a specific size and consisted of 126 utilities.

Table 8-8 shows how costs for the index groups were correlated to CMP's actual historical costs.

The data showed a reasonably strong correlation between the index groups and CMP when viewed over the long run. From this perspective, a little more than 90% of the variation in CMP's non-fuel costs per customer can be correlated with changes in the non-fuel costs per customer for the large and small samples.[13] The annual cost changes for the large group were correlated with only 9% of the historical changes in CMP's non-fuel costs. The small group performed even worse, with only a 3% correlation.

The statistical correlations were adequate to form the basis of a CMP proposed indexing plan. The specific CMP proposal was not

[13] A causal relationship between the prices at CMP and those of the comparative groups is not proposed. These results suggest that 90% of the change in bills at CMP may be explained by factors that resulted in bill changes at the comparable utilities.

adopted, but the analysis contributed to the revenue-per-customer decoupling plan described in Chapter 4.

Bill Index Conclusions

In conclusion, customer bills are *not* a perfect yardstick of least-cost planning performance. Bill indexing may, however, be better than the current alternatives which often neglect supply-side options and only involve incentives to foster DSM investments based on shared savings approaches. We have also demonstrated that one of the limits, the exclusion of customer contributions from customer bills, may *improve* the overall incentive features of a bill-indexing approach by creating a reasonable balance between the reduction of overall resource costs, on the one hand, and the price impacts on non-participants in DSM programs on the other. In our analysis average customer bills are always lower when utilities successfully pursue least-cost planning and the bill impact is the same for equally cost-effective DSM or supply-side resources.

Every system of regulation will create gaming opportunities and potentially perverse incentives. Any regulatory approach inevitably produces examples of inappropriate behavior that could enrich the utility or produce other negative consequences. For example, under bill indexing: utilities might seek to defer costs to a future period, preferring to see average bills lowered in the short-run at the expense of long-run bill impacts; utilities might discourage customers from using more electricity even when it may be cost-effective; or utilities may encourage customers to invest in non-cost-effective DSM options, thereby causing consumption and average customer bills to decrease.

Incentive plans are not intended to operate as a complete substitute for least-cost planning. The "correct price signals" theoretically existing in the unregulated competitive market are also not a substitute for securities laws and enforcement. Least-cost planning will necessarily include regulatory oversight and public review and participation in the planning process. There simply is no perfect regulatory system, nor are there perfect markets. There are only better systems and worse systems.

We conclude our review of some of the features of various types of bill indices by noting that:

• Using a bill index to provide incentives for least-cost planning does not seem to present any serious internal inconsistencies or conflicts with least-cost planning;

- Using a bill index on an annual basis does not appear to be as effective or statistically sound as the use of a cumulative index approach over an extended period of time; and

- External bill indexing offers the advantage of creating desirable incentives for utilities to manage all factors affecting customer bills, namely price and quantity. Despite some weaknesses, an econometrically based external bill index approach may be most equitable for a broad range of utilities. Because it adjusts for key differences among utilities, it may be easier to adapt to a wide range of circumstances than the composite approach.

- Internal bill indexing is generally more limited in scope, but may pose fewer implementation problems than external bill indexing. The simplest and most circumscribed approach uses statistical techniques to measure the impact of DSM programs on average customer bills. The need to adjust for factors such as weather makes the comparison to a utility's own least-cost plan slightly more complex.

Acknowledgments

Funding for this work came from the New York State Energy Research and Development Authority, the United States Department of Energy, and the New York Department of Public Service.

References

California Public Utilities Commission and California Energy Commission. 1987. *Standard Practice Manual for Economic Analysis of Demand-Side Management Programs*. San Francisco, Calif.

Lowry, Mark, Laurits Christensen, and Douglas Caves. 1991. *Final Report on External Bill Indexing*. Madison, Wis.: Christensen Associates.

Moskovitz, David. 1988. "Will Least-Cost Planning Work Without Significant Regulatory Reform?" In National Association of Regulatory Utility Commissioners' *National Conference on Least-Cost Planning*. Aspen, Colo.

National Association of Public Utility Commissioners (NARUC). 1988. *Least-Cost Utility Planning: A Handbook for Public Utility Commissioners, Volume II*. December.

Chapter 9

Utility Energy Services

Charles J. Cicchetti and Ellen K. Moran

Introduction

Environmental consciousness has recently pushed electric utilities into the conservation debate. Utilities are increasingly being asked to pursue energy conservation and load management programs, under the rubric of demand-side management (DSM), to reduce the wasteful use of resources. However, the disincentives of DSM programs (among which are disallowance of program costs, lost revenues resulting from reduced sales, and erosion of rate base) are greater than the incentives to many utilities. Investments in DSM can reduce a utility's profitability. As businesses that are accountable to shareholders, investor-owned utilities are reluctant to undertake actions that have a negative effect on their bottom line. The disincentives to utility-sponsored conservation, however, can be overcome through an intelligent reworking of the traditional rate-of-return regulatory scheme. Based on the premise that an appropriate goal of public policy is to maximize the amount of cost-effective utility-sponsored conservation, coupled with the belief that economic incentives are preferable to command-and-control regulation to promote economic efficiency, utilities and state commissions have developed incentive regulation plans for DSM programs. Although these plans vary in scope, structure, and detail, they share the common objective of transforming DSM into a potentially profitable activity for utilities, by including some or all of the following components:

- Recovery of program costs;

- Compensation for lost revenue (more precisely, lost profits); and

- A positive financial incentive for the utility.

The various disincentives and measures to overcome them and provide inducements to utilities to undertake DSM programs have been presented elsewhere in this book. This chapter focuses on an emerging approach, the energy services concept or energy service charge (ESC), and its prospects for encouraging utility investment in conservation.

The Energy Services Concept

Much of the conceptual thinking about current energy services programs comes from "Including Unbundled Demand-Side Options in Electric Utility Bidding Programs" (Cicchetti and Hogan 1989). The Cicchetti-Hogan proposal was designed to facilitate including demand-side programs in bidding systems. A utility would purchase conserved kilowatt-hours (kWhs) from a customer/supplier in an all-source bid program only if the cost per kWh of energy savings is less than the marginal cost of new supply-side alternatives. Customers who deliver conserved kWhs would then receive conservation services and would continue to pay for all kWhs of conservation-related energy services as part of the regular bill for utility service.

The energy services concept distinguishes between a utility's sale of energy commodities (Btus and kWhs) and its sale of energy services. As applied to electric utilities, the concept is based on the idea that the demand for kWhs is an indirect, as opposed to direct, demand. Thus, the demand for kWhs is derived from the demand for the underlying services that they provide for consumers, such as heating, cooling and lighting. Conservation can produce energy services in the same way as kWh sales. For example, if a utility installs a conservation measure at a customer's premises that leaves the consumer as comfortable as before, or with the same end-use services, but at an annual savings of 100 kWhs, then the installed measure has produced energy services equivalent to the 100 kWhs saved.

Energy services may be an amorphous concept that is difficult to define and measure. Nonetheless, energy conservation programs necessarily include some estimate of the kWhs the customer will save by using a particular conservation technique. This engineering estimate of savings may both define and quantify the energy service provided by the utility. Under an energy services regulatory scheme, the customer pays for this service as part of the regular bill for utility service. In the simplest case, the estimated kWhs saved would be added to the actual consumption of electricity. The customer would pay for the energy service provided at a price up to the full retail value of the saved kWhs, as well as for the electricity provided, priced at the standard retail rate for kWhs.

Suppose, for example, that a conservation device that leaves the

Table 9-1. Utility Demand/Supply Options

Retail rate ($/kWh)	.05
Incremental supply cost ($/kWh)	.06
Conservation cost ($/kWh)	.05
Initial load (kWh)	2,000
Incremental/saved load (kWh)	100

	Before	Supply Option	Traditional Demand Option	Demand Option with Full ESC	Demand Option with Incentive ESC
Revenues	100	105	100	105	104
Costs	90	96	95	95	95
Profits	10	9	5	10	9

customer at the same level of comfort can be installed for 5¢/kWh or less; incremental generation costs 6¢/kWh; and the retail energy rate is 5¢/kWh. If the utility installs the conservation measure charging the customer 5¢/kWh saved, total system costs are lowered compared to the supply alternative, the consumer is just as well off as before, and the utility's earnings have remained the same or increased. Unlike the more traditional DSM approach, the kWhs of energy services will be added to sales, but not to generation. This is the key differentiating feature of the energy services concept. As in the traditional DSM approach, the costs of conservation are added to the other fixed and variable costs of the utility. Alternatively, the utility could offer the customer an incentive of up to 1¢/kWh saved (a net energy service charge of 4¢/kWh) and still remain as profitable as if it had elected the supply option.

In general, if the cheapest resource addition is conservation (consistent with the principles of least-cost planning) and if the one-for-one exchange of electricity for conservation energy services leaves the customer as well off as before, then utility profits will not suffer as they do under the traditional approach to DSM accounting. As shown in Table 9-1, they may remain the same. Compared to the supply option, profits will increase by the difference between the cost of conservation and the cost of the best alternative resource, less any incentive payment to promote customer participation.

The following sections present some detail about how the energy services program treats participants and non-participants, how it could be used in bidding programs and some other advantages.

Participants Versus Nonparticipants

Utility-financed energy efficiency programs are a paradox to utilities, customers and regulators. Although energy efficiency measures may be less expensive than supply-side resources, their successful implementation can cause electricity prices to increase. Utilities rely on customer payments to meet the fixed expenses of existing generating facilities, transmission lines, distribution facilities, and customer services networks. If some customers, by taking advantage of utility-sponsored DSM programs, reduce their purchases and their utility bills, average prices must rise for all remaining sales to recover the same amount of fixed costs and to cover the added DSM program costs. Nonparticipants in the utility DSM program, including those who have already invested in energy efficiency measures, can see their rates rise without receiving any direct benefit.

The energy service charge has been proposed to respond to these equity concerns. A conservation program with the energy service charge collected from program participants protects nonparticipants from rate increases resulting from conservation investments made by utilities that benefit selected customers. Through the energy service charge, participating customers pay for the investments made directly on their behalf while still enjoying lower overall energy costs due to kWh savings. Tables 9-2 and 9-3 present comparative analyses of various supply and demand options from the perspectives of the sponsoring utility, the customer actively participating in the DSM program, and the nonparticipant customer. In these tables the two conservation strategies, the traditional utility-funded program and the energy service charge approach, have deliberately been presented as lower cost options than the supply scenario (and, therefore, superior from a societal perspective for meeting additional load), when utility marginal costs are both above (Table 9-2) and below (Table 9-3) average prices.

Under most scenarios the participating customer is better off with DSM than in either the initial state or the supply case. This may be so even if the customer pays for the conserved kWhs. The opposite is true for the nonparticipant who is at a distinct disadvantage under a traditional demand-side program when conserved kWhs are not paid for by the direct beneficiaries. The cost burden of the program is shifted in part to nonparticipants who enjoy none of the energy services provided

Table 9-2. DSM Pricing Examples / Marginal Costs Above Average Price				
Initial load (000 kWh)				100
Incremental/saved load (000 kWh)				10
Initial embedded cost ($/kWh)				.06
Incremental supply cost ($/kWh)				.09
DSM measure cost ($/kWh)				.08
	Initial State	Supply Option	Traditional Demand Option	Demand Option with Full ESC
Utility Perspective				
Revenue requirement ($)	6,000	6,900	6,800	6,800
Generated electricity (000 kWh)	100	110	100	100
Units sold (000 kWh)	100	110	100	110
Average rate ($/kWh)	.06	.063	.068	.062
Customer Perspective				
Participant:				
Conventional load (000 kWh)	12	12	10	10
Conservation savings (000 kWh)	0	0	2	2
Units purchased (000 kWh)	12	12	10	12
Total charge ($)	720	753	680	742
Nonparticipant:				
Conventional load (000 kWh)	12	12	12	12
Conservation savings (000 kWh)	0	0	0	0
Units purchased (000 kWh)	12	12	12	12
Total charge ($)	720	753	816	742

by the conservation. Although conservation represents the least-cost alternative, the nonparticipant bill under this scenario is driven higher than both in the initial state and under the supply-side option. Those participants who benefit from "free" conservation do so at the expense of all ratepayers, especially those ratepayers who choose not to participate.

Table 9-3. DSM Pricing Examples / Marginal Costs Below Average Price				
Initial load (000 kWh)				100
Incremental/saved load (000 kWh)				10
Initial embedded cost ($/kWh)				.06
Incremental supply cost ($/kWh)				.05
DSM measure cost ($/kWh)				.04
	Initial State	Supply Option	Traditional Demand Option	Demand Option with Full ESC
Utility Perspective				
Revenue requirement ($)	6,000	6,500	6,400	6,400
Generated electricity (000 kWh)	100	110	100	100
Units sold (000 kWh)	100	110	100	110
Average rate ($/kWh)	.06	.059	.064	.058
Customer Perspective				
Participant:				
Conventional load (000 kWh)	12	12	10	10
Conservation savings (000 kWh)	0	0	2	2
Units purchased (000 kWh)	12	12	10	12
Total charge ($)	720	709	640	698
Nonparticipant:				
Conventional load (000 kWh)	12	12	12	12
Conservation savings (000 kWh)	0	0	0	0
Units purchased (000 kWh)	12	12	12	12
Total charge ($)	720	709	768	698

Under the energy services approach, there is a more equitable distribution of program costs and benefits between participants and nonparticipants. When marginal costs are below average prices, the energy service option leads to the lowest rates for all consumers. When marginal costs are above average prices, only the initial state offers lower rates to all ratepayers.

Energy Services in Bidding Programs

An advantage of the energy services concept is that it provides the proper signals for demand-side bidding and facilitates the inclusion of demand-side options in all-source bidding programs. Admittedly, the complexities of comparing options with different load shape impacts, timing, and reliability make it difficult to construct scoring systems that can accurately compare supply-side options to demand-side options. But these are second-order technical problems compared to the issue of price signals when a utility is comparing two seemingly similar offers: a supply-side offer to deliver a kWh of electricity at a certain time and place in exchange for a payment by the utility (P_s), and a demand-side offer to save an identical kWh of electricity in exchange for a payment by the utility (P_D). It would appear that the utility should award the bid to the party offering kWhs at the lowest price, P_s or P_D, and that the utility should be indifferent, *ceteris paribus*, whether the winning bidder represents the supply-side or demand-side option.

However, this cost-effectiveness comparison is incomplete; such partial analysis may lead the utility to purchase excessive quantities of saved kWhs. This is because the amount P_D is not the only money that changes hands if the utility accepts the demand-side bid. The demand-side bidder (whether a third-party provider or host facility) receives the direct payment P_D from the utility. In addition, the metered customer on whose premises the measures are installed gets to keep the money it otherwise would have paid the utility for the kWh it no longer needs to buy. The savings are an indirect payment (P_R) from the utility to the demand-side bidder and represent lost revenues (associated with the fixed cost portion of rates) from the utility perspective. The proper comparison then becomes P_S versus $P_D + P_R$. But if the energy services concept has already been put into place (effectively eliminating or substantially reducing P_R), a demand-side bid could, under those circumstances, be compared directly to a supply-side bid. Pursuant to the energy services program, winning demand-side bidders would continue to pay the retail price for the kWhs they are saving (P_R). With the energy service charge paid by the customer-as-consumer, the danger of double payment (i.e., the bid price paid by the utility plus the benefit of a reduced utility bill) to the customer-as-supplier is removed as is the consequent overinvestment in conservation.

Other Advantages of Energy Services

While the energy services concept distributes benefits fairly, eliminates cross-subsidization and alleviates upward rate pressure, it also offers a number of other, less critical advantages as a DSM approach. It minimizes several forms of gaming behavior to which other mechanisms may be susceptible. For example, failure on the part of customers to follow through with the installation of committed measures can be readily detected through the inspection process by the utility. Besides, by having to pay for both kWhs consumed and those purportedly saved, there is a real disincentive for nonperformance. At the same time, the utility has an incentive to minimize costs (as opposed to maximizing dollars invested in DSM or kWh saved) and to install only cost-effective measures since its opportunity to make money rests on the difference between program costs and the energy service charge collected.

Two utilities have moved beyond theory to practice in terms of implementing a variation of the energy services concept. Following are descriptions of these two programs (descriptions of PacifiCorp and Bangor Hydro are based on September 1991 information).

PacifiCorp's Energy Service Program

In 1989 PacifiCorp Electric Operations (PacifiCorp), the merged system of Pacific Power and Utah Power, had a load of 4,861 megawatts (MWs). Under the first combined resource plan for the utility system, this load was projected to grow to 6,557 MWs by 2008. Despite a substantial energy surplus for most of the 1980s, deficits are projected for the forthcoming 20-year planning horizon. As part of its least-cost planning process, the company has identified achievable conservation potential system-wide of 400 to 600 MWs that would entail gross utility investment over the planning horizon of $1.3 billion in 1989$ (Pacific Power and Utah Power 1989).

Utility customers often do not undertake investments for energy improvements even when such investments can lower their total costs over the lifetime of the investments. The barriers that customers face include capital constraints, high implicit discount rates, short payback requirements, uncertainties, relatively low utility bills in proportion to total budgeted expenditures, lack of adequate information, or inconvenience (Bhattacharjee, Cicchetti, and Rankin 1991). Traditionally, PacifiCorp sought to overcome these impediments with programs that offered generous financial incentives to customers. Throughout the 1980s they pioneered a number of programs that sought maximum market penetration by paying customers the full cost of DSM measures. Through direct experience the utility has witnessed the upward pressure on rates stemming from such traditional rebate programs and the mismatch of costs and benefits between program participants and nonparticipants. The company is now adopting the energy service approach to overcome the market barriers and redress imperfections in its delivery system.

In Oregon PacifiCorp, doing business as Pacific Power & Light Company (PP&L), introduced the Energy Services Program which was approved by the Oregon Public Utility Commission (OPUC) in June 1990. It includes two components: the first is directed statewide at new commercial building construction; the second is currently an experimental program to upgrade energy efficiency in existing commercial buildings in a single city. For new commercial buildings the utility offers building owners an energy analysis of the proposed structure during the design phase to identify opportunities for improved energy efficiency beyond the measures mandated by building codes. Based on the audit, PP&L will pay for design assistance, materials, and installation of cost-effective energy conservation measures in exchange for the right to share in energy savings under an energy services contract.

For this program, energy conservation measures are defined as permanently installed structural improvements that can reduce overall electric energy use (Pacific Power & Light Company 1990). The measures include a variety of heating, ventilating, air-conditioning, lighting, and control options, with more than half the potential savings attributable to lighting applications. They offer greater comfort and convenience, higher quality of lighting, better aesthetics, more control, and higher productivity and typically achieve 10% to 30% percent more energy efficiency than can be attained through code compliance. One of the chief purposes of this particular program is to capture what would otherwise become lost opportunities. Installing energy-efficient measures in new construction is far easier and more cost-effective than retrofitting an existing structure with the same measures.

For each possible measure, there is a cost-effectiveness calculation or measure funding limit that is determined by multiplying the measure's estimated annual kWh savings (beyond mere adherence to code) by 110% of PacifiCorp's levelized avoided cost (levelized over the expected life of the measure), derived from its most recent least-cost plan. The 10% premium or "conservation adder" is intended to give conservation measures a cost advantage vis-à-vis generation resources and is dictated by the Pacific Northwest Electric Power Planning and Conservation Act of 1980.

The utility and building owner then enter into a contract, specifying the measures to be installed and estimating the energy service charge payable over the contract term in monthly installments. The contract term is selected by the owner to be the lesser of the weighted average life of the installed energy conservation measures or 20 years. The energy service charge is tied to market interest rates: the prime rate for cost-effective measures and prime plus 3% for measures for which costs exceed the measure funding limit but which are requested by the owner. However, the charge is set in such a way that participants are assured at least a 5% discount from applicable retail electric rates (PacifiCorp Electric Operations 1991). The energy service charge is structured like a loan payment for tax purposes. Unless treated as a loan from the utility, the conservation payments made by the utility (the design assistance, capital and installation costs) would be ruled as taxable income for the participants.

When the energy conservation measures have been installed, inspected and monitored, a one-time adjustment in favor of the customer may be made. If the post-installation energy audit shows that the energy savings actually achieved are less than projected, there is a quantity reduction in the formula for the charge. On the other hand, if the savings are greater than projected, the customer retains all of the benefit.

Although the building owner enters into the contract with the utility, it is the building occupant receiving metered electric service who actually pays the energy service charge which appears as a separate line item on the monthly utility bill. However, if the customer fails to pay, the utility can charge the building owner. At the end of the contract term, the charge is removed from the bill and the customer then receives any remaining benefits at no additional cost. At any time during the contract, the energy service charge can be eliminated from the bill by a termination or buy-out payment made by either the owner or the customer in an amount that is the remaining net present value of the contract.

PacifiCorp is in the process of introducing similar programs aimed at the new commercial construction sector in the other states (California, Idaho and Utah) where its operating divisions provide electricity service. The company is also planning to introduce pilot energy service programs to its industrial and residential customers.

What is the likely impact of the program on the utility and its customers? In Table 9-4, the assumptions and middle column, "Hypothetical Ideal Case," describe a scenario presented by PacifiCorp in its May 1990 concept paper on the energy services program (PacifiCorp Electric Operations 1990). The case shows that there is a relatively even sharing of program benefits between the utility and its customers. Two alternative scenarios are shown that confirm the program's sensitivity to variations in installed costs, the impact of the conservation adder, and the level of discount or sharing available to participants. If the utility could install DSM measures more cheaply (as in Case I), it could increase profits or it could afford to give prospective participants a deeper discount below retail rates. On the other hand, if the customer chooses a package of measures priced at the cost-effectiveness ceiling or measure funding limit, the utility will face greater capital costs to implement the program. In Case III, there is only a small margin between average retail price and marginal cost (part of which is eroded by the mandated conservation adder), and the utility loses money by proceeding. By reducing the customer discount (and incentive) for energy services or by eliminating the conservation adder, the utility can be made indifferent in economic terms.

The energy services concept represents a new approach to energy conservation. Based on an early assessment of PP&L's experience in Oregon, the most attractive feature to building owners appears to be the availability of 100% financing on flexible terms for capital improvements that enhance asset value and improve operating costs. The ability to confirm actual savings is important, as is the willingness of PP&L to absorb risk on energy savings. The concept appears to be producing reasonable savings at a low cost to the utility. Customers are receptive to the program in its current form and it appears capable of reaching high penetration rates (McDonald 1991).

Both PacifiCorp and the OPUC staff recognize the program's novelty and acknowledge that many of the assumptions regarding its economics are unsubstantiated. As a result, the company has been directed by the OPUC to comprehensively assess the program after its first full year of operation. As part of the assessment, the company will evaluate whether the program is capturing energy conservation opportunities that might otherwise be lost, whether the program maintains revenue neutrality between supply- and demand-side alternatives, and whether the program is achieving its goal of leaving all players (utility, participating and nonparticipating customers) better off (Pacific Environments 1990). To test some of the underlying assumptions, the company will examine the accuracy of estimated savings during the design phase compared to adjusted savings estimates, the level of energy savings realized, the degree to which recommended measures are implemented, the level of market penetration achieved, and overall cost-effectiveness of the program. The company will also attempt to profile characteristics of participants and nonparticipants and to survey trade allies (architects, engineers, contractors) as to program awareness and attitudes.

Bangor Hydro's Energy Service Program

Another DSM program based on the energy service concept is Bangor Hydro-Electric Company's (BHE) Payload Program. Under this program, BHE would become an energy service company offering two substitute goods, kWh service and DSM service, directly and through a customer bidding program (Bangor Hydro-Electric Company 1989a).

Under the bid program, qualified applicants (either building owners or energy service companies) would offer for sale kWs and/or kWhs resulting in load shifting from the utility's peak and/or improved efficiency of existing end-uses. The company planned to evaluate bids on the basis of price (which had to be less than the total avoided cost per energy or demand unit conserved or shifted), feasibility and reliability, expected life of the project, measurability of load impacts, economic stability of the sponsor's business over the life of the project measures, and the bidder's ability to actually construct and manage the project. Upon approval of an application, BHE would pay the participant a lump sum equal to the amount bid for the project. In exchange, the successful bidder would enter into a contract with BHE agreeing to make periodic payments in an amount equal to the contracted energy conserved or peak demand reduced, at a per-unit rate competitive with the customer's applicable billing rate over a period equal to the estimated useful life of the installed measures (Bangor Hydro-Electric

Table 9-4. Energy Service Charge / Financial Illustration	
ASSUMPTIONS	
Average retail price ($/kWh)	0.050
Embedded and marginal cost ($k/Wh)	0.045
Discount rate—utility (nominal)	10.0%
Consumption before efficiency measures (kWh/year)	15,000
Expected kWh savings per year	4,000
Life of measures (years)	20
Utility's cost of conservation ($)	1,200[2]
($/year, nominal levelized)	141[2]
($/kWh)	.035[2]
Energy service charge @ 90% of retail ($/year)	180

Table 9-4 (continued)			
IMPACTS	**($/year)**		
	I At Average Cost for Achievable Savings[1]	**II Hypothetical Ideal Case[2]**	**III At Cost-Effectiveness Ceiling[3]**
Customer Perspective			
Without program:			
Electric bill	$750	$750	$750
With program:			
Electric bill	$550	$550	$550
Energy service charge	$180	$180	$180
Total cost	$730	$730	$730
Savings from program	$20	$20	$20
Company Perspective			
Without program:			
Electricity sales			
Revenue	$750	$750	$750
Cost of power	$675	$675	$675
Net revenue	$75	$75	$75
With program:			
Electricity sales			
Revenue	$550	$550	$550
Cost of power	$495	$495	$495
Net revenue	$55	$55	$55
Energy service charge impact			
Energy service charge revenues	$180	$180	$180
Cost of efficiency measures	$120	$141	$198
Net energy service charge revenues	$60	$39	($18)
Total net revenue	$115	$94	$37
Benefit from Program	$40	$19	($38)

[1] Based on average cost per achievable kWh of savings for new commercial buildings, Pacific Division (2.8¢/kWh in 1988$, adjusted for inflation), reported in Table 3, page 22, in the OPUC and ODOE December 1989 report, "An Assessment of the Conservation Activities of Pacific Power & Light, Portland General Electric, and Idaho Power Company."

[2] Ideal case as presented by PacifiCorp Electric Operations in May 1990 report, "Concept Paper: Energy Service Approach to Improving Customer Energy Efficiency."

[3] Maximum cost of conservation at assumed marginal cost plus 10% conservation adder, as per test of cost effectiveness (.045 + .0045 = .0495).

Company 1989b). While this program has been proposed, an RFP released, and bids received, to date no contracts have been executed and recent developments suggest that the program may be shelved or reconsidered.

BHE has been involved in extended proceedings with the Maine Public Utilities Commission (MPUC) about its record in designing and implementing DSM programs, the outcome of which has ramifications for its energy service program. The parties hold divergent philosophical views on issues central to the whole DSM debate (Maine Public Utilities Commission 1990a). For example, BHE has consistently maintained that the appropriate test of cost-effectiveness for DSM is one in which the utility pays no more than the difference between its marginal cost and the average price for conservation programs. Given that its marginal cost is well below its retail rates, the computed number is negative for BHE, leading its management to conclude that it should not purchase conservation even if it is less costly than avoided supply. The company's stated goal in designing conservation programs has been to minimize program costs to reduce the absolute effect on rates and to minimize cross-subsidization of program costs by nonparticipants (Linnell 1991). The company also says that the consumer has adequate incentives to directly invest in DSM due to savings in bills. Consistent with this view, BHE has tried to design DSM programs that induce customer participation by providing low-cost information and educational materials, that arrange access to capital, and that ensure that participating customers pay for benefits received through the program (Linnell 1990).

The MPUC, on the other hand, has established that the appropriate test of DSM cost-effectiveness is the "all ratepayers" or total resource cost test. In Chapter 380 of the public utilities code, the breakeven point for this test is defined as the conservation investment in which program costs equal program benefits. Program costs include the utility's costs related to the DSM program (excluding revenue losses) plus the total of all participants' costs (the incremental costs incurred to save electricity). In choosing among DSM programs, utilities are encouraged to give priority to programs with the greatest net present value under the total resource cost test. For high-priority programs that nonetheless adversely impact rates (causing average revenue requirements per kWh to increase), a utility is instructed to give priority to programs that are the most widely available and that distribute benefits to as many customer classes as possible.

Consequently, the Commission has recently adjusted BHE's authorized rate of return on equity downward by 50 basis points due to the company's inadequate performance in the areas of least-cost planning and DSM (Maine Public Utilities Commission 1991). In so doing, the Commission found that by applying a more restrictive test of cost-effectiveness than that set out in Chapter 380, not only had BHE failed to comply with the Commission's rulings and the Maine Energy Policy Act, but it had underinvested in DSM measures and unduly delayed implementation of programs. The Commission went on to state that "inefficiencies in the conservation and demand-side management area hurt ratepayers both in terms of higher present energy bills and in terms of lost opportunities for least-cost planning." In closing, the Commission concluded that the "company needs to institute new policies, new standards, new procedures, new practices, and new plans that would ensure that new goals can be achieved in the most efficient way under a correct concept of what is cost-effective" (Maine Public Utilities Commission 1991). In its new mood of compliance, BHE is now re-examining all of its DSM programs, including the energy service program which never really got off the ground.

The debate between BHE and MPUC has been presented at length to illustrate one of the drawbacks of the energy services concept. In a jurisdiction such as Maine that relies on the total resource cost test, utilities and their regulators are unlikely to embrace the energy services approach. The approach is best suited for jurisdictions where passing the no-losers' test remains an important component of the overall test of cost-effectiveness. As with other approaches to DSM, the appeal of the energy services concept largely depends on the standards by which it is evaluated.

Is There an Incentive?

While it is clear that the energy services approach removes many of the disincentives to the achievement of energy efficiency, a question remains as to whether the concept provides utilities with a positive incentive to pursue DSM programs aggressively. The OPUC has described the energy service program as a "pioneering approach to energy conservation" that will "increase market penetration of new energy efficiency technologies by giving PP&L a profit incentive" (Oregon Public Utility Commission 1990). However, in recommending the program to the full Commission, the OPUC staff said that the program's philosophy was the recoupment of program costs, including lost revenues, from program participants. No mention was made of an incentive for the utility (Combs 1990). The OPUC staff also classified the energy service program as a method of cost recovery, as opposed to a method to recover lost profits or to provide a positive incentive (Hellman, Hagerman, and Busch 1990). This apparent confusion or divergence of opinions certainly gives cause to question what exactly is being accomplished by the energy service charge. How much of a financial incentive is there that contributes to the utility bottom line and benefits shareholders? The aggregate energy service charge collected from participating customers will be treated by PacifiCorp as an offset against revenue requirements to minimize the rate impacts on nonparticipating customers. Consistent with the view that DSM measures are capital-intensive investments made for future service (as an alternative to supply-side investments), the utility is able to add conservation investments to its rate base. Pursuant to an OPUC order, all eligible conservation program expenditures, including promotional activities, concessions, acquisition costs and capital costs, are subject to deferral and amortization from the date placed in service over the useful life of the assets. The utility is thus able to receive its normal rate of return on the unamortized portion of the conservation investment remaining in its rate base.

There are a few exceptions to this general accounting treatment for conservation-related expenses. Conservation advertising costs are to be expensed in the year incurred and kept separate from other advertising costs. Nonetheless, the Commission left the door open for possible future deferral for DSM program advertising costs. The costs of certain legislatively mandated programs are also expensed, as are ongoing operational costs, e.g., those for program reporting and tracking, as with supply-side alternatives.

An analysis of the current ratemaking treatment granted PP&L for its energy service program in Oregon leads to the conclusion that the program does not offer an explicit financial incentive. By offsetting energy service tariff receipts against revenue requirements, the mechanism provides only for cost recovery. There is no makeup of lost revenues. The company recognizes that while the impact of annual lost net revenues may be low in the near-term, the cumulative effects over time could become substantial, without subsequent changes in ratemaking treatment. The OPUC staff has proposed treating net revenue losses as an appropriate acquisition cost eligible for special accounting treatment (Hellman, Hagerman, and Busch 1990; *Electric Utility Week* 1991b). While this step has not yet been taken on behalf of PacifiCorp, Portland General Electric was recently permitted recovery of lost revenues (*Electric Utility Week* 1991b). This action offers the prospect of more rewarding ratemaking treatment for PacifiCorp's energy service program in the future.

Perhaps the removal of some disincentives is sufficient to promote DSM. But not all risks have been fully eliminated. In collecting the energy service charge over a period of up to twenty years, PacifiCorp still bears considerable risk of disallowance over time. The regulatory risk increases proportionally with the length of time over which the cost recovery is spread. Because the utility wants to avoid rate hikes, this risk of disallowance is the price to be paid in the trade-off between expensing and amortizing DSM costs. Despite the absence of positive financial incentives, PacifiCorp is still a strong advocate of the energy service concept because it makes good sense and is consistent with its least-cost plan objective to stabilize rates.

However, in California where PacifiCorp operates in a few counties, it has been granted more favorable ratemaking treatment for its energy service program. It is able to retain the first year's energy service revenues as an incentive for the account of shareholders. The California Public Utilities Commission (CPUC) discontinued the company's electric revenue adjustment mechanism (ERAM), at PacifiCorp's request, and replaced it with the energy service charge (California Public Utilities Commission 1990).

As part of that decision, the CPUC approved three levels of accounting treatment for DSM program expenditures. The first level is for direct costs of energy-efficiency measures, which are amortized over the energy service tariff term and receive the company's allowed rate of return on the unamortized portion. The second level is for expenditures that lead to the purchase of conservation resources, such as those for staff labor, field implementation, and program development. These are recovered in revenues over a one-year period, carry a 5% incentive, and are also subject to a budget limit and spending ratio of expenses to DSM assets. The third level is for all remaining costs that are expensed. In addition to the explicit first-year incentive and three-tiered cost recovery approach, the CPUC has also set minimum performance targets for DSM programs. Failure to meet the minimum standards, on a program-by-program basis, will result in an after-the-fact 50% reduction in the rate of return on the 1991–1993 expenditures for each deficient program and a corresponding reduction in earnings on expenditures over the remaining life of the particular program. The California approach, with its penalty and rewards, represents a balanced and economically efficient solution to the challenge of incentive ratemaking.

With respect to ratemaking treatment for BHE's DSM activities, the MPUC treats conservation costs in accordance with Chapter 37 of the public utilities code (Maine Public Utilities Commission 1990a). Briefly, Chapter 37 sets out rules for the recovery of costs of utility-funded conservation programs and establishes an energy conservation adjustment clause. Eligible costs include reasonable costs of purchase of goods and services, financial subsidies for conservation loan programs, and other market research, advertising, and promotional costs. The MPUC may direct that costs of measures having a useful life of five years or more be recovered over more than one year and may make a reconciliation adjustment to the energy conservation adjustment for such reasons as under- or overcollection and imprudent or unreasonable energy conservation procurement practices. The Commission may also grant an incentive adjustment, either positive or negative, up to 10% of the cost of each program during the current energy conservation adjustment period.

Referring back to the three elements of a reformed ratemaking scheme to transform conservation into a rewarding business for utilities (recovery of program costs, compensation for lost revenue, and a positive financial incentive), Chapter 37 addresses the first of these components, overlooks the second and offers the prospect of the third. It is regrettable that the long-running adversarial proceeding, involving BHE's recalcitrance and the MPUC's resort to punishment, does not create a climate conducive to reworking the traditional rate-of-return regulatory environment in favor of motivational management.

Energy Service Program Complexities and Disadvantages

The energy services program, like most DSM programs, does have its share of disadvantages. One of the perceived drawbacks of the energy service concept is that the customer receives no commodity of real value. It appears that the customer is asked to pay for electricity that is not consumed. If the customer were truly paying something for nothing, energy services would be a hard policy to recommend or explain. However, if the conservation improvements are real, then the customer is truly providing an energy service. If the customer wishes to be paid for providing that service, then it should also pay for the energy retained. By using a combination of engineering estimates and post-installation audits, the customer can be shown that the savings are not phantom, but represent a reduction in energy consumption which will provide a real benefit over time.

As with many DSM mechanisms, the energy service concept is likely to be most effective when the difference between the cost of the demand-side resource and the best available supply-side alternative is large, since the maximum inducement that the utility can pay the participant will vary directly with this difference. A problem may occur when both conservation and generation costs are substantially below the retail energy rate. Suppose that the marginal cost of conservation is 2¢/kWh, the marginal cost of generation is 3¢/kWh, and the retail energy rate is 5¢/kWh. A customer who is unwilling to conserve on his own for a net benefit of 3¢/kWh will be unlikely to accept the utility's offer to participate in its conservation program if the maximum incentive payment is 1¢/kWh (the difference in the utility's cost for the two resource options). Utility customers who are not inclined to participate in traditional DSM programs may be even less willing to do so under an energy service concept that requires them to pay their own way. Customers accustomed to offers of "free" conservation and "double benefits" will find unattractive the obligation to pay for DSM improvements in their utility bills over long periods of time. Only an intensive consumer education effort regarding utility ratemaking and pricing issues might overcome this obstacle.

PacifiCorp dealt with a number of program complexities that became apparent in the process of implementing a new idea. A few examples follow that are not necessarily unique to the energy services concept.

The payments prescribed by the energy service tariff are the obligation of the customer receiving electricity service during the term of the energy services contract. At the same time, the utility has recourse to the building owner for any energy service charges that the customer fails to pay. A question arises when the property is sold. To protect its long-term interests, PacifiCorp requires the seller to assign the energy services contract to the buyer at the time of transfer of the property. To further protect itself, the utility is permitted by the terms of the tariff to record the contract as an encumbrance (but not a lien) against the property.

Changes in the tenancy of a building over a twenty-year contract term may lead new tenants to question the energy service charge on their utility bill. It may be difficult to convince new occupants that they benefit from prior installations with utility bills lower than what they would otherwise be. While PacifiCorp has taken steps to ensure that successor customers will be responsible for the energy service charge, the utility sought additional protection in the form of state legislation which was enacted in April 1991. The purpose of the legislation is to clarify that the OPUC has the authority to require that successor customers continue to pay the contracted charges (Griffith 1991). With the adoption of the legislation, PacifiCorp may drop the assignment requirement which can unduly complicate property sales and hinder program acceptance. The enactment of this legislation was certainly not a prerequisite to program implementation, but it will serve to enhance program acceptance, particularly as PacifiCorp attempts to extend energy service programs to its industrial and residential customers.

One way to compare DSM programs is in terms of their ease of administration. The energy service program developed by PacifiCorp would receive only a fair rating in this category. A package of DSM measures tailored for each site and accompanied by an energy audit following installation entails substantial front-end supervision and transaction costs. Booking and tracking a portfolio of loans, each with its specific payment terms (interest rate, monthly payment amount, and term based on the useful life of the measures), will require significant accounting systems and resources, as will monitoring changes in occupancy and ownership. The cost per transaction may decline over time as the program grows and as operating systems are developed.

Conclusion

The energy services concept is relatively untried, compared to the other DSM incentive approaches described in this book. It is just now emerging from concept to implementation. Indeed, the evolution of the energy services program at PacifiCorp and the adaptations that the company has made over the past year demonstrate the importance of utility responsiveness to field experience and market research results.

While there still may be much to learn through full-scale implementation of the energy services program, it has a number of distinctive features that set it apart from other DSM programs. First and foremost is the process of "unbundling" energy services, which recognizes the value of conservation as a product marketed separately from the sale of electricity. Introduction of an energy services program puts a utility squarely in the mode of marketing and selling conservation to its customers. Through the energy service charge, the product of energy efficiency stands out as a separate line item on the bill for the customer to see and evaluate. The relationship between product, value received, and price is clearly evident, which is not always the case for other utility-sponsored conservation programs. Under traditional approaches, individual participants may see a reduction in consumption but they fail to recognize that the benefit is accompanied by a general upward pressure on rates for all customers.

The energy services program requires that participants accept the long-term benefit of substantial savings that are realized only after measures have been paid for in full. This feature may prove to be the program's undoing if it hinders widespread acceptance by customers. If that happens, utilities may be hesitant to adopt the approach for fear that it will compromise their achievement of conservation targets established with their regulators. Experience to date is limited; only time will tell how successful the energy services concept will actually be.

Conservation programs that are funded through general rate increases have been criticized as sources of cross-subsidization between customer classes. Customers who are already energy-efficient or who do not receive benefits from DSM programs have been critical of these approaches. The energy services program responds to these criticisms by more equitably allocating costs and benefits. Participants receive a high-value energy service, lower electricity bills, verification of the realized savings, and financing for the whole enterprise. Non-participants enjoy lower and more stable future electricity prices because participants implement DSM measures that postpone the need for new generating resources. The utility receives revenue that offsets the cost of the DSM investment and reduces the need for price increases, while acquiring a cost-effective, long-lived electric resource.

While the energy service charge does not provide an incentive mechanism, it does recover program costs and makes up the contribution to fixed charges for the foregone electricity sales. Only as part of an overall package of ratemaking mechanisms is the energy service charge likely to provide an incentive to the sponsoring utility.

There are those who question why positive financial incentives are needed to induce utility executives to aggressively pursue DSM programs, particularly if cost recovery and makeup of lost revenues are assured. In response, perhaps we need to ask why utility decisionmakers would expend the substantial level of effort required to embark upon an unfamiliar and innovative marketing endeavor, with an uncertain outcome, only to be no better off than they were before. The prospect of a reward challenges utility management to apply their entrepreneurial skills and creativity to identify and pursue DSM opportunities. A properly structured incentive based on program outputs or value created, as opposed to program inputs, will further the policy objective of maximizing cost-effective conservation. A well-structured and significant incentive can mean the difference between success and failure in a utility's acquisition of demand-side management as a meaningful electric resource.

References

Arrington, ZoeAnne, Lee Sparling, Katherine Beale, Charles Grist, John Savage, and Charles Stephens. 1989. *An Assessment of the Conservation Activities of Pacific Power and Light, Portland Gen-*

eral Electric, and Idaho Power Company. Oregon Public Utility
Commission and Oregon Department of Energy.
Bangor Hydro-Electric Company. 1989a. *Least-Cost Planning Perfor-
mance Proposal of Bangor Hydro-Electric Company.*
Bangor Hydro-Electric Company. 1989b. Filing with Maine Public
Utilities Commission re: *Bangor Hydro-Electric Company's Pro-
posed Conservation and Load Management Customer Proposal
Solicitation Program Filed Pursuant to Chapter 380.*
Bhattacharjee, Vinayak, Charles J. Cicchetti, and William F. Rankin.
1991. "Are There Any Economic Efficiency Arguments for
Embracing Utility-Sponsored Conservation Programs?" Paper
presented 1 July at Western Economic Association 66th Annual
Conference, in Seattle, Wash.
California Public Utilities Commission. 1990. *Order Instituting Inves-
tigation of Pacific Power and Light Company Electric Rates and
Charges for Electric Service.* Decision 90-12-022.
Cicchetti, Charles J., and Suellen M. Curkendall. 1988. "Conserva-
tion Subsidies: The Economists' Perspective." *Electrical Potential*
2(3). Pp. 3–12.
Cicchetti, Charles J., and William Hogan. 1989. "Including Unbun-
dled Demand-Side Options in Electric Utility Bidding Programs."
Public Utilities Fortnightly 123(12). Pp. 9–20.
Cicchetti, Charles J., and Rod Shaughnessy. 1980. "Is There a Free
Lunch in the Northwest? (Utility-Sponsored Energy Conservation
Programs)." *Public Utilities Fortnightly.* Pp. 11–15.
Combs, Barbara. 1990. *Pacific Power and Light Advice* No. 90-101.
Oregon Public Utility Commission Staff Report.
Electric Utility Week. 1991a. "Pacificorp Plan Seeks to Overcome Bar-
rier to Conservation Investment."
Electric Utility Week. 1991b. "Oregon PUC Staff Wants Utilities
Reimbursed for Sales Lost to DSM." August. Pp. 11–12.
Griffith, Bill. 1991. Testimony to Oregon Senate Business Housing
and Finance Committee re: *Senate Bill 864.*
Hellman, Marc, James Hagerman, and Ed Busch. 1990. "Investiga-
tion into Incentives for Acquisition of Conservation Resources by
Electric Utilities." Oregon Public Utility Commission. Draft.
Katz, Myron. 1989. "Utility Conservation Incentives: Everyone
Wins." *The Electricity Journal.* October. Pp. 26–35.
Linnell, Stephen. 1990. "Proposed Increase in Rates." Rebuttal Testi-
mony in Docket No. 90-001 before Maine Public Utilities Com-
mission.
Linnell, Stephen. 1991. "Investigation into Bangor Hydro-Electric
Company's Performance, Policies and Management Practices in

the Areas of Demand-Side Management and Integrated Least Cost Planning." Testimony in Docket No. 90-286 before Maine Public Utilities Commission.

Maine Public Utilities Commission. 1990a. "Bangor Hydro-Electric Company Proposed Increase in Rates." Stipulation and Order Approving Stipulation in Docket No. 90-001.

Maine Public Utilities Commission. 1990b. "Investigation into Bangor Hydro-Electric Company's Performance, Policies and Management Practices in Areas of Demand-Side Management and Integrated Least Cost Planning." Supplemental Notice of Investigation and Procedural Order in Docket No. 90-286.

Maine Public Utilities Commission. 1991. "Investigation into Bangor Hydro-Electric Company's Performance, Policies and Management Practices in the Areas of Demand-Side Management and Integrated Least-Cost Planning." Decision in Docket No. 90-286.

McDonald, Gordon. 1991. "Energy FinAnswer: Program for New Commercial Buildings." Presentation by PacifiCorp Electric Operations to Washington Utilities and Transportation Commission Staff.

McNamee, William A., and Wayne L. Lash. 1990. "The Potential Use of Competitive Bidding for Resource Acquisition by Investor-Owned Electric Utilities." Oregon Public Utility Commission and Oregon Department of Energy.

Oregon Public Utility Commission. 1989. "In the Matter of the Application of Pacific Power and Light Company for Reauthorization of Deferred Accounting for its Model Conservation Standards Program." Order No. 89-1700.

Oregon Public Utility Commission. 1990. "PUC Lauds PP&L's Conservation Program as an Oregon 'First.'" Press Release.

Pacific Environments. 1990. "Energy Efficiency Program Evaluation Plan for Existing Commercial Building and Industrial Facilities." Pacific Power and Light Company.

PacifiCorp and Pacific Power and Light Company. 1988. "Application Under the Public Utilities Code of the State of California for Authority to Eliminate the Electric Revenue Adjustment Mechanism (ERAM)." Application A88-10-014.

PacifiCorp Electric Operations. 1990. "Concept Paper: Energy Service Approach to Improving Customer Energy Efficiency."

PacifiCorp Electric Operations. 1991. Advice No. 91-1 to Idaho Public Utility Commission re: Commercial Energy Services, Schedule 120.

Pacific Power and Light Company. 1990. Advice No. 90-100-Revised to Oregon Public Utility Commission.

Pacific Power and Utah Power. 1989. "Planning for Stable Growth: Resource and Market Planning Program."

SBW Consulting, Inc. 1990. Analysis of Commercial Model Conservation Standards Study, Final Report. SBW Consulting Report No. 9001, submitted to Bonneville Power Administration.

Whittaker, Curtis M. 1988. "Conservation and Unregulated Utility Profits: Redefining the Conservation Market." *Public Utilities Fortnightly* 122(1). July 18–22.

Chapter 10

Evaluation of DSM Programs and Financial Incentives

Eric Hirst

Introduction

Will utility demand-side management (DSM) programs fulfill their promise to provide large amounts of cost-effective and environmentally benign energy and capacity resources? Utilities and regulatory commissions will consider DSM programs credible resources only if their performance is measured accurately.

As the number, size, scope, cost, and effects of DSM programs increase, evaluations will become increasingly important. New challenges will face evaluators in the 1990s; a key challenge is the role evaluation will play in determining the amount of financial incentive a utility earns for its DSM programs. Because of the importance of evaluations, state public utility commissions (PUCs) will pay close attention to evaluation methods and results.

Most regulatory incentives for DSM programs are tied to estimates of the benefits these programs provide. Figure 10-1 shows a typical shared-savings mechanism. In this scheme, the total benefit of a DSM program is based on the estimated reductions in electricity use and demand (GWh and MW) multiplied by the appropriate avoided costs (energy or capacity). The net benefit is the difference between total benefit and program cost. PUCs typically award the utility a share of this net benefit.

However, the utility often must meet minimum performance requirements before it receives any benefit (See Chapter 6). In the Figure 10-1 example, if the utility fails to achieve 40% of the target net benefit, shareholders lose money. If it achieves from 40% to 60% of the target, shareholders neither win nor lose. If the net benefit exceeds 60% of the target, the utility earns 38% of the net benefit over the 60%

threshold. Thus, if the utility meets its target ($1.3 million in this example), its incentive of $0.19 million is 15% of the total net benefit.

Evaluations determine the total benefits of DSM programs. This chapter begins by explaining what evaluation is, why it is important for electric utilities, and the key steps in evaluation. The next section then discusses engineering analysis, a frequently used approach to estimate program energy and load reductions. The fourth section presents a hypothetical example of problems that might arise when evaluations are subject to litigation, a likely occurrence when utility earnings depend on evaluation results. A hypothetical, rather than real, illustration is used because shared-savings incentives are so new that none has yet been tested with evaluation results in regulatory proceedings. The next section suggests possible resolutions to the problems discussed in the previous section. The sixth section mentions other issues related to evaluations and financial incentives, and the final section presents conclusions.

What Is Evaluation?

Evaluation, the systematic measurement of the operation and performance of DSM programs (Hirst 1990),[1] relies on objective measurements rather than anecdotal evidence or personal impressions. Evaluations use social science research methods and technical data to ensure valid results. Evaluations are intended to influence future decisions about DSM programs; they are not academic exercises.

Evaluations provide information for program managers and staff to improve program operations and for utility executives and regulatory agencies to assess these programs. For example, evaluations can:

- Document the energy savings, load reductions, and cost-effectiveness of DSM programs;

- Show ways to improve programs by increasing participation rates, raising energy savings, or cutting costs;

- Suggest ways to improve the design of future DSM programs;

- Support DSM budgets before the utility's budget committee; and

- Provide data to strengthen the company's load forecasts and resource planning.

---

[1] See Energy Program Evaluation Conference (1991) and Keating and Hicks (1990) for examples of evaluations of conservation and load-management programs. See Hirst and Reed (1992) for discussions of the key issues when evaluating DSM programs.

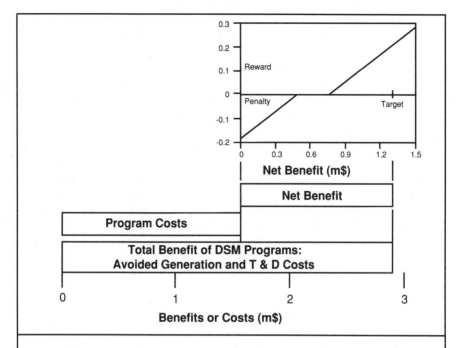

Figure 10-1. Utility Keeps Fraction of Net DSM Benefit in Shared-Savings Approach
Schematic showing the mechanics of a shared-savings mechanism to reward utility shareholders for implementing cost-effective DSM programs. In this example the utility projects program costs at $1.6 million, the total benefits at $2.9 million, and the net benefit at $1.3 million. Shareholders will earn $0.19 million if the utility achieves its net-benefit target of $1.3 million.

Utilities should measure DSM program performance using the same competence and diligence used to monitor power plant performance. Utilities have detailed information for each of their power plants on construction costs and time; operations, maintenance, and fuel costs; heat rate, availability factor, and capacity factor; the duration and causes of each outage; and fuel consumption, plant output, and emissions. Unfortunately, comparable data do not exist for utility energy-efficiency and load-management programs that describe program participation rates, energy savings (GWh) and load reductions (MW), and program costs.

There are two types of evaluations. *Process* evaluations examine program operations to identify how well the program is implemented and to suggest ways to improve it. These evaluations focus on program goals, history, and activities, and are based largely on interviews with utility staff, managers, participants, and trade allies associated with the program.

Impact evaluations examine the program's effect, by providing quantitative documentation of benefits and costs. Impact evaluations measure participation, participant acceptance of the recommended DSM measures and practices, performance of the DSM technologies being promoted, energy and load reductions, and costs.

Impact evaluations compare what happened to program participants with what would have happened to them if the program had not existed. These evaluations deal with two types of energy savings and load reductions. *Total* savings represent the participants' reductions in annual electricity use and peak demand. *Net* savings represent that portion of the total savings that can be directly attributed to the utility program. Thus, net savings are the difference between total savings and the savings that participants would have made if the program had not existed. Nonprogram savings reflect customer responses to changes in electricity and fossil-fuel prices, changes in economic activity or personal income, introduction of new electricity-using technologies, and other nonprogram factors. Energy savings and load reductions are determined primarily from analysis of monthly electricity bills and load-research data; these electricity-use data are often supplemented with data on weather, occupant and operating characteristics, and facility characteristics.

Careful evaluations can transform guesses, estimates, numbers, and data into useful information on the costs, performance, and operations of DSM programs. Program evaluation is integral to responsible management, and is as important for DSM programs as for power plants. Evaluations are especially important—and likely to be controversial—when money is transferred from utility customers to shareholders on the basis of evaluation results. PUCs and other interested groups (e.g., environmental and consumer organizations) must be confident that the net benefits shared between customers and shareholders truly exist! Utility planners also require credible information on the performance of DSM programs to ensure system reliability and provision of low-cost electricity services to customers.

Use of Engineering Approaches to Estimate Savings

The simplest, most widely used, and least expensive evaluation method uses engineering estimates of energy savings and load reductions. For simple DSM measures, such as compact-fluorescent exit signs in commercial buildings, electricity savings and load reductions are estimated using simple calculations. For more complicated measures, such as computerized control systems in commercial buildings, the estimates are based on sophisticated heat-loss models that simulate the energy flows into and throughout a building.

Unfortunately, current engineering estimates are often incorrect and inappropriate, because actual savings are often less than the engineering estimates (Hirst et al. 1985; Greely et al. 1990; Nadel and Keating 1991). The engineering calculations often (though not always):

• Use incorrect assumptions concerning operating and maintenance practices (e.g., hours of use and temperature settings) and neglect interactions among DSM measures (e.g., the effect of changes in lighting loads on air-conditioning and heating loads);

• Ignore changes in occupant behavior induced by the DSM measures installed (e.g., increases in indoor temperatures or decreases in wood use for heating after retrofit); and

• Do not account for quality-control problems in selecting and installing DSM measures.

Engineering estimates are inadequate alone because they generally do not reflect differences between net and total savings. This occurs because the engineering calculations are usually applied only to participant facilities.

Moskovitz (Chapter 1) notes that reliance on engineering estimates gives perverse incentives to utilities for their DSM programs. For example, if savings from a program are 700 MWh, as calculated with engineering estimates, but actual energy savings are 800 MWh, then the utility loses money because sales are 100 MWh lower than expected. Conversely, if actual savings are only 600 MWh, then the utility profits because sales are greater than expected. Furthermore, in states that allow recovery of lost revenues caused by DSM programs, the utility profits in two ways—once on the revenues associated with the extra 100 MWh of sales, and once on the recovery of lost revenues based on an assumed 600 MWh savings even though not all of these revenues were lost.

Thus, competent evaluations are important to determine the amount of net lost revenue caused by the utility's DSM programs and to determine the amount of financial incentive to give utility shareholders. If the utility in this hypothetical example operates with a revenue decoupling mechanism, such as the Electric Revenue Adjustment Mechanism used in California, then the problems associated with inappropriate recovery of net lost revenue do not exist (Chapter 3). However, the perverse incentive exists for shared-savings mechanisms.

Engineering estimates might be appropriate to use as the basis of the utility incentive where program benefits are small and measurement costs are high (e.g., water-heater wraps). In such cases, the engineering estimates of electricity savings could be based on conservative assumptions. Such low-cost programs are expected to be very cost-effective; thus, use of conservative assumptions will still leave the program cost-effective.

Bench testing the technologies promoted by the program and measuring the number of installations, including the number of measures removed and the number of participants who would have installed the measures without the program, should still be done. Expensive end-use metering would not be used in these cases because metering costs could exceed program benefits. For example, Pacific Power & Light (PP&L) conducted a small experiment on low-flow showerheads (Delta T Inc. 1989). The one-month, $15,000 project included 553 phone calls to PP&L customers, of whom 11% agreed to have the company install a low-flow showerhead. Installers measured the flow rates for the old and new showerheads. Customer satisfaction and showerhead retention rates were measured with a post-installation telephone survey. Results from this experiment, while not statistically valid, provide useful information for simple engineering calculations of energy savings.

As utilities begin DSM programs, engineering estimates may be a useful and noncontroversial way to design the initial stages of an incentive system. The California Collaborative (1990), which included utilities, government agencies, and other groups, agreed to an incentive system based on prior engineering estimates of savings for individual DSM measures. These estimates will be revised on the basis of evaluations, but only after the programs (and the associated incentives) have been in place for three years.

As discussed later, utilities can overcome the limitations of engineering analysis by combining engineering calculations with more rigorous evaluation methods and data in order to refine the assumptions used in engineering estimates. Given time and experience, engineering estimates can be improved to provide inexpensive, rapid, and reasonably accurate estimates of program performance.

Evaluation in Contested Hearings

More and more PUCs are providing financial incentives for utilities to implement cost-effective DSM programs (Chapter 2). The most popular incentives use shared-savings mechanisms, in which the utility keeps part of the net benefit provided by its DSM programs (Figure 10-1). Shared-savings mechanisms are popular because they encourage the utility to minimize costs and to maximize the net benefit. The critical element in computing net benefits is estimating energy and demand reductions, the province of program evaluators.

Consider a hypothetical commercial lighting program as an example of the ambiguities in a carefully conducted evaluation. (I leave to the reader's imagination the controversies that might arise over a poorly designed and run evaluation.) This program targets office buildings and includes general information and on-site lighting audits. These activities identify suitable lighting measures and encourage the customer to apply for the 50% rebate offered by the utility. The rebate helps to defray the costs of purchasing and installing energy-efficient lamps, ballasts, fixtures, and controls.

In this hypothetical example, the utility's comprehensive evaluation included three elements (see Violette at al. [1991] and Xenergy [1990] for discussions of these and other evaluation approaches):

• Analysis of two years of electricity-billing data, one year before and after participation, for samples of participants and eligible nonparticipants;

• Thirty days of time-of-use metering, pre- and post-retrofit, of a sample of lighting circuits for a sample of participants only (no comparison group); and

• Engineering analysis of the energy and load reductions caused by the measures actually installed by participants (again, no comparison group).

The utility used multiple methods to estimate program savings because each method is imperfect. If the utility plan shows how the results of these disparate methods will be used to determine program effects, such triangulation can build confidence in the estimates ultimately used. Not surprisingly, these approaches and their associated analytical procedures produced different estimates of energy savings (Table 10-1); for simplicity the load-reduction effects of the program are ignored in this example.

Comparison of pre- and post-retrofit electricity use, based on monthly billing data, for participants showed a reduction of 9,600 kWh/year. Nonparticipants showed an increase in electricity use over the same two-year period, leading to a net savings of 12,800 kWh/year. The local economy was growing during this period, which led to higher occupancy levels and longer hours of operation for these buildings. These changes in building use help explain the increase in electricity use for nonparticipants. The customers who received a lighting audit but did not apply for the rebate (one-third of the customers that received audits) also cut their consumption. Table 10-1 shows the roughly 50% difference in estimated electricity savings based on analysis of billing data. These differences depend on whether or not rebate-participant savings are adjusted for nonparticipant changes in electricity use and for the savings achieved by audit-only participants.

The data from short-term metering showed savings roughly comparable to those from analysis of the billing data for the participants only. Complications arose in scaling up the metering results to a full year. The metering covered roughly 60 days over a three-month period, with the middle month devoted to installing new lighting measures. The amount of electricity used for lighting varies with season, and is larger in the winter than in the summer. Conversely, the indirect electricity savings associated with reduced air-conditioning loads are greater in the summer than in the winter.

There were similar problems in estimating the savings with engineering calculations. These calculations are based on the change in connected load multiplied by the number of hours of use per year. The assumed change in connected load did not, however, include the fact that many of the new lamps replaced ones that were burned out. Thus, the estimated reduction in load, and therefore in electricity use, was too high.

The percentage differences in electricity savings among these approaches are magnified when estimating net benefit (Table 10-1), because net benefit is the difference between total benefit (directly proportional to energy savings) and program cost. In this example, the program cost (including the customer contribution to the cost of the retrofit measures) averaged $1,630 per rebate participant. Table 10-1 shows estimates of net benefits, based on an avoided cost of 6¢/kWh and a measure lifetime of four years. Program costs are thus roughly half the total benefit. So, a 10% error in estimating total benefit leads to a 20% error in net benefit.

As shown in Figure 10-1, the utility incentive is usually a share of the net benefit, adjusted for a minimum threshold level. In this example, the target net benefit was based on a planned savings of 12,000 kWh/participant, 1000 rebate customers, and a per-participant cost of $1,630, which yields a net benefit of $1,250 per participant. If the threshold is 60% of the target value, then the utility receives an incentive only if the net benefit exceeds $750 per rebate participant. Again, following Figure 10-1, the incentive gives the utility shareholders 15% of the net benefit if the utility achieves its target net benefit; the incen-

Table 10-1. Effects of Evaluation Results on Estimates of DSM-Program Net Benefits and Utility Incentive

Evaluation Method	Results per Rebate Customer			
	Energy Savings (kWh/year)	Gross Benefit[a] ($)	Net Benefit[b] ($)	Utility Incentive[c] ($)
Bill analysis				
Rebate participants only	9,600	2300	670	0
With comparison group	12,800	3070	1,440	260
+ audit only[d]	15,200	3650	2,020	480
Ad hoc metering	10,040	2410	780	10
Engineering analysis	14,400	3460	1,830	410

[a] Gross benefit = energy savings × $0.24/kWh, where $0.24 is the value of $0.06/kWh avoided costs over the four-year average life of the measures.

[b] Net benefit = gross benefit − $1,630, where $1,630 is the cost per rebate customer.

[c] Utility incentive = 0.38 × (net benefit − $753) if net benefit > $753; otherwise incentive = 0. In this calculation, $750 = 60% × $1,255 is the savings per customer target.

[d] Savings/rebate-participant = 12,800 + (0.5 × 4,800), where 0.5 derives from the 2:1 ratio of rebate recipients to audit-only recipients, and 4,800 kWh is the net savings achieved by audit-only recipients.

tive fraction is 38% of the net benefit above the 60% threshold level (Table 10-1). Thus, the structure of the incentive mechanism adds more leveraging in going from net benefits to utility incentive (Figure 10-2).

The variations in estimates of program savings in this hypothetical example are typical of those found in actual evaluations. Tonn and White (1990) used monthly electricity billing data to estimate the effects of the Model Conservation Standards in multifamily buildings in Tacoma, Washington. Their estimates differed according to the sophistication of the models used to explain household electricity use and according to whether the analysis dealt with dwelling units (apartments) or apartment buildings (Table 10-2).

Possible Resolutions

Which of the results in Table 10-1 is correct? Perhaps more important, which estimate should the PUC use to determine the incentive paid to the utility? Several approaches are possible that differ in who does the evaluation and how it is planned and conducted.

The utility and PUC (and perhaps other parties, such as interven-

Table 10-2. Estimates of Space-Heating Electricity Use and Savings (kWh/ft²-year) in New Multifamily Buildings in Tacoma, Washington

	Model Conservation Standard	1983 Current Practice	Difference
Dwelling-unit analysis			
Simple means	8.3	7.1	−1.2
One equation	7.5	9.8	2.3
Four equations	7.2	9.2	2.0
Seven equations	6.9	8.7	1.8
Building-level analysis			
Simple means	4.1	5.5	1.4
One equation	4.7	5.5	0.8
Source: Tonn and White (1990).			

ors) could agree before the program starts on the evaluation methods to use, including sample frames and sizes, data-quality controls, and analysis methods. Prior specification of evaluation methods may be unworkable because programs evolve as they are implemented. Indeed, one of the major strengths of DSM resources compared to supply resources is their small unit size and flexibility. Predetermining the specifics of an evaluation would rob the program of important flexibility. Also evaluation is as much art as science; it is impossible to predetermine the appropriate actions to take and criteria to use throughout the evaluation. However, by agreeing beforehand on the data sets and methods to use in identifying net savings, some of the problems cited in the preceding section could be avoided.

The utility and PUC could select an independent group to conduct the evaluation. The evaluation contractor would conduct the evaluation and recommend an estimate of program energy and load impacts to the PUC. This approach is also problematic. First, evaluations should be closely coupled to other customer-analysis activities within the utility. Contracting out the evaluation prevents the utility from gaining valuable analytical experience and data on its customers, although contractors can be required to provide data to the utility. Such an approach limits development of in-house utility staff and infrastructure. Finally, using an independent contractor to conduct the evaluation does not, by itself, eliminate controversy from the PUC hearing. The utility, PUC

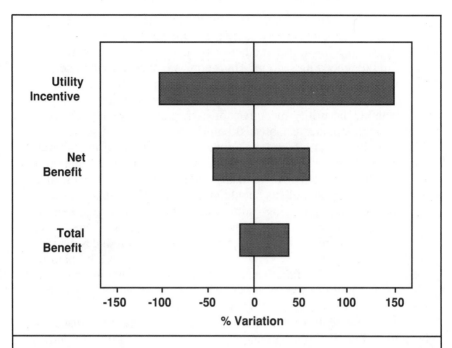

Figure 10-2. Shareholder Incentive Leveraged by Evaluation Results
Approximate leveraging of evaluation results in going from estimates of total benefits to net benefits and to utility incentive, for a hypothetical lighting program.

staff, and intervenors are still free to suggest alternative interpretations of the evaluation.

Using a collaborative process to design the evaluation and to review its progress might reduce controversy. In this approach, the utility conducts the evaluation, but shares the planning and oversight with the PUC and other organizations. California and Massachusetts use this approach. In California, the PUC Division of Ratepayer Advocates hired a contractor to review utility evaluation plans, progress, and results. In Massachusetts, the Conservation Law Foundation is working with utilities throughout the evaluation process.

Finally, the commission could appoint an independent expert to review the utility's evaluation approach and results. Based on this review, the expert would suggest to the commission estimates to use in computing the benefits of the utility's DSM programs. This situation occurred in Wisconsin, when the consultant suggested revisions to the engineering estimates developed by the utility (Nichols et al. 1990).

Figure 10-3 illustrates another possible resolution of the dilemma raised in the preceding section. This approach, similar to the one used in California, employs both engineering estimates and evaluations based on measured electricity use, and involves nonutility parties. In this method, the utility incentive for each program year is based on engineering estimates established before the start of that year. These estimates, in turn, are based on competent evaluations of the program during the previous year. The arrows from program to evaluation to engineering estimates involve many judgments. Therefore utilities, commission staff, and others should jointly interpret evaluation results and develop the engineering estimates for the following year's incentive. This approach partly decouples the utility incentive from evaluation results, but only one year at a time. This method reduces the risks for utilities, because they know how their incentives will be paid for the coming year. This iterative approach should yield engineering estimates that rapidly converge to the estimates based on analysis of billing data, load-research data, and survey results.

To what extent are program evaluations along the lines discussed here affecting PUC determination of financial incentives for utility DSM programs? Unfortunately, it is too soon to know. For example, New England Electric (1991) filed its first evaluation report with the Massachusetts Department of Public Utilities in June 1991. In October 1991, the Department provisionally approved the results and the associated incentive for the company. The incentive will be adjusted in late 1992 based on the company's evaluation report filed in June 1992. In New York and California, the Commissions are seeking to adjust the engineering estimates developed by the utilities. Evaluations using billing data or load-research data have not been an important part of a utility's application for financial incentives in either state.

Other Issues

This paper focuses primarily on the role of evaluations in determining the energy savings and load reductions generated by utility DSM programs. Evaluations affect (and are affected by) the design and operation of DSM programs and their incentive mechanisms in other ways.

Determining the cost of a DSM program, although conceptually more straightforward than determining energy savings, is also an important function of evaluations. For incentive mechanisms that are based on total resource costs, utilities must measure customer contributions to the cost of installing measures promoted by the program.

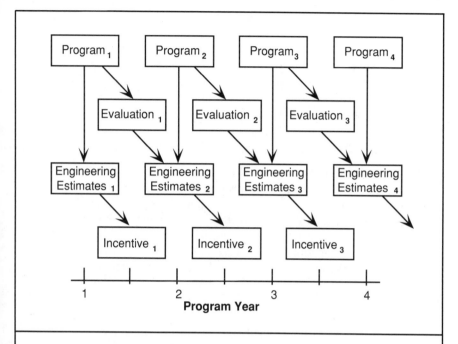

Figure 10-3. Use of Evaluations and Engineering Estimates to Compute Financial Incentives
Suggested iterative approach for estimating energy and load reductions for DSM programs. Subscripts refer to the year of program operation. (Evaluations often require a year or more to complete; thus the link between evaluation results and engineering estimates might take two years.)

Berry (1989) discusses the various costs associated with DSM programs.

A more subtle issue concerns utility and PUC selection of programs that can be readily evaluated. Because it is extremely difficult to measure the effects of information-only programs, these programs may be dropped even if they are inexpensive. Similarly, programs targeted at trade allies may not be conducted because of difficulties in their evaluation. Consider, for example, a program that encourages appliance dealers to stock and promote energy-efficient units by having utility staff pose as potential purchasers for random visits to dealers. If the dealer promotes energy-efficient units, a cash prize is awarded. Such a program, while potentially cost-effective, is much harder to evaluate than a program that offers rebates to customers who buy such appliances. In California, this dilemma was resolved by paying incentives for such programs on the basis of utility expenditures rather than on the basis of evaluation results.

These and other evaluation issues are not explored here for two reasons. First, the primary role of evaluations, and the area in which most controversy will likely occur, is estimating energy and load impacts. Second, in some cases even less is known about these other issues than about the energy and load effects.

Conclusions

Serious evaluation of utility DSM programs is a recent activity and the use of evaluation results in determining utility earnings is just starting. Therefore, the kinds of problems that might arise in PUC hearings and possible resolutions are unknown. It is clear, however, that competent evaluations of utility DSM programs will be increasingly important (Weil 1990). For example, New England Electric has seven full-time professionals plus an annual contract budget of almost $3 million for evaluating of DSM programs (Destribats et al. 1991). Puget Power (1991), after running ambitious DSM programs for more than a decade, prepared its first evaluation plan, largely because of regulatory reforms that remove disincentives and offer incentives for DSM programs.

If incentives for DSM programs are to succeed, indeed, if DSM programs are to become legitimate energy and capacity resources, then evaluations will become much more important. The number and quality of evaluation staff, both in utilities and PUCs, will increase. Also, program evaluation approaches will slowly become standardized. Utilities and commissions will not use engineering analysis as the primary basis for awarding financial incentives. Integrated approaches to evaluation will evolve that rely primarily on electricity-use data (monthly bills and load-research data) combined with analytical methods and the judgments of utilities, commissions, and other interested parties.

The Maine Public Utilities Commission (1991), in its order on incentive mechanisms:

> includes a system for measuring the savings achieved by successful conservation and load management efforts . . . [which] . . . requires rigorous analysis of a sample of customer bills before and after the installation of efficiency measures and a comparison of these changes with a control group of customers who did not participate in the conservation program.

This statement describes a model for other commissions and utilities to use as they develop evaluation approaches for their financial incentive mechanisms.

Acknowledgments

I appreciate the many helpful comments on a draft of this paper from Harley Barnes, Gregory Haddow, Elizabeth Hicks, Kenneth Keating, Michael Messenger, David Moskovitz, Dan Quigley, Don Schultz, Gary Swofford, and the editors of this book. This work was sponsored by the Office of Conservation and Renewable Energy, U.S. Department of Energy under contract with Martin Marietta Energy Systems, Inc.

References

Berry, L. 1989. *The Administrative Costs of Energy Conservation Programs*. ORNL/CON-294. Oak Ridge, Tenn.: Oak Ridge National Laboratory.

California Collaborative. 1990. *An Energy-Efficiency Blueprint for California, Appendix A: Measurement Protocols for DSM Programs Eligible for Shareholder Incentives*. Report of the Statewide Collaborative Process, available from California Public Utilities Commission. San Francisco, Calif.

Delta T Inc. and Solar Energy Association of Oregon. 1989. *Product Placement Report of Low Flow Showerheads for Pacific Power and Light*. Portland, Oreg.

Destribats, A.F. et al. April 1991. "Demand-Side Management at New England Electric: Implementation, Evaluation, and Incentives." *Proceedings of the National Conference on Integrated Resource Planning*. National Association of Regulatory Utility Commissioners, Washington, D.C. Pp. 108-115.

Energy Program Evaluation Conference. 1991. *"Energy Program Evaluation" 1991 International Energy Program Evaluation Conference*. CONF-910807. Chicago, Ill.

Greely, K., J. Harris, and A. Hatcher. 1990. *Measured Savings and Cost Effectiveness of Conservation Retrofits in Commercial Buildings*. LBL-27568. Berkeley, Calif.: Lawrence Berkeley Laboratory.

Hirst, E. 1990. "Evaluating Demand-Side Management Programs." *Electric Perspectives* 14(6). Pp. 24–30.

Hirst, E., D. White, E. Holub, and R. Goeltz. 1985. *Actual Electricity Savings for Homes Retrofit by the BPA Residential Weatherization Program*. ORNL/CON-185. Oak Ridge, Tenn.: Oak Ridge National Laboratory.

Hirst, E., and J. Reed, eds. 1992. *Handbook on Evaluation of Utility*

DSM Programs. ORNL/CON-336. Oak Ridge, Tenn.: Oak Ridge National Laboratory.

Keating, K., and E. Hicks, eds. 1990. "Program Evaluation." *ACEEE 1990 Summer Study on Energy Efficiency in Buildings.* Vol. 6, *Program Evaluation.* Washington, D.C.: American Council for an Energy-Efficient Economy.

Maine Public Utilities Commission. 1991. "Investigation of Chapter 382 Filing of Central Maine Power Company." Docket No. 90-085. Augusta, Maine.

Nadel, S., and K. Keating. 1991. "Engineering Estimates vs Impact Evaluation Results: How Do They Compare and Why?." *In Energy Program Evaluation: Uses, Methods, and Results, 1991 International Energy Program Evaluation Conference.* CONF-910807. Chicago, Ill. Pp. 24–33.

New England Electric System. 1991. *1990 DSM Performance Measurement Report.* Westborough, Mass.

Nichols, D., G. Katz, D. Singh, N. Talbot, and E. Titus. 1990. *Savings from the Smart Money Program, An Audit of the 125 Mega Watt Demand Reduction.* Boston, Mass.: The Tellus Institute.

Puget Power. 1991. *Demand-Side Management Measurement and Evaluation Plan.* Bellevue, Wash.

Tonn, B., and D. White. 1990. "Energy Savings in New, Low-Rise Multifamily Buildings Due to Energy-Efficient Building Practices." *Energy Systems and Policy* 14(2). Pp. 85–111.

Violette, D., M. Ozog, M. Keneipp, F. Stern, and P. Hanser. 1991. *Impact Evaluation of Demand-Side Management Programs.* Vol. 1, *A Guide to Current Practice.* CU-7179. Palo Alto, Calif.: Electric Power Research Institute.

Weil, S. 1990. "The Urgent Need for Verifying DSM Achievements." In *ACEEE 1990 Summer Study on Energy Efficiency in Buildings.* Vol. 6, *Program Evaluation.* Washington, D.C.: American Council for an Energy-Efficient Economy. Pp. 6.215–6.223.

Xenergy Inc. 1990. *Demand-Side Management Program Evaluation Scoping Study.* ESEERCO Project EP90-34. New York: Empire State Electric Energy Research Corp.

Chapter 11

Spare the Stick and Spoil the Carrot: Why DSM Incentives for Utility Stockholders Aren't Necessary

Paul Newman, Steven Kihm, and David Schoengold

Introduction

All state regulatory commissions are not created equal, nor are all commissions necessarily interested in being equal. Each commission establishes procedures and philosophies to create its own identity, consistent with the laws of each state. Because procedures and philosophies differ from commission to commission, the need for incentives for demand-side management (DSM), and the types of DSM incentives that may be appropriate, will vary from commission to commission.

In this chapter we first look closely at how regulation is practiced by the Wisconsin Public Service Commission (Wisconsin PSC). We then examine several DSM incentive mechanisms that have been tried by the Wisconsin PSC, before considering arguments for incentives in light of financial theory and empirical data on returns to utility investors. Our analyses support the Wisconsin PSC's conclusion that DSM incentives for utility stockholders are unnecessary. In our view, if commissions elsewhere adopted Wisconsin's regulatory practices and policies, stockholder incentives for DSM could be eliminated.

Wisconsin's Regulatory Practices

Annual Rate Relief

Every major electric and gas utility in Wisconsin is required to file annually for rate relief. This procedure was implemented in the early

1980s primarily for administrative reasons; however, the implications for DSM have been profound.

Annual rate review does *not* lead to extra work for the Wisconsin PSC staff. In fact, it probably has led to more efficient staff use. Complete hearings for the major utilities last two weeks at most, with some ending in two or three days. The frequency of the cases allows for both quick and thorough reviews. For example, with annual rate review, auditors are more familiar with the utility and its current operations, and therefore the audit can be completed faster. It is easier for staff to find the key issues quickly as they are familiar with the recent cases and already know the important issues in each case. Also, having worked on several other rate cases each year, analysts can quickly incorporate new issues discovered in one case into other cases whenever appropriate.

Another factor that makes the process work is that each utility's filing date is staggered to allow staff to move from case to case. For example, Wisconsin Power and Light Company's hearing is scheduled for June each year, while Wisconsin Electric Power Company's is scheduled in October. So, at the beginning of the year everybody knows how many rate cases there will be and when they will occur. Some would question whether such procedures could be adopted in states with many large utilities. In Wisconsin, this procedure is used for the eight largest electric and gas utilities without major scheduling problems.

Future Test Year

In each annual rate case, the Wisconsin PSC uses a future (i.e., fully forecasted) test year. This prevents financial problems for a utility that plans to significantly increase its demand-side spending, because rates can be set in anticipation of the future DSM budget. In contrast, jurisdictions that use a historic test year set rates on the basis of historic DSM spending levels, which will be lower if spending is increasing; the result is a disincentive to increase DSM spending.

The future test year, coupled with annual rate review, has made recovery of lost revenues a minor or even non-existent issue in Wisconsin. The effects of demand-side programs are considered when forecasting test year sales levels. So, if demand-side programs meet their targets, there will be no lost revenues. On the other hand, if the utility promotes either more or less demand-side activity than was projected when setting rates, the revenue shortfall or windfall will be short-lived. In the next annual rate case, the actual results of the demand-side programs will be considered when setting rates. While this process does not guarantee dollar-for-dollar recovery of lost revenues, it has worked well enough to prevent any noticeable lost revenue problems.

Accounting for DSM Expenditures

Two specific DSM accounting procedures are part of the Wisconsin PSC's standard ratemaking approach: escrow accounting for DSM expenses and capitalization of certain DSM costs.

In many jurisdictions, utilities are concerned they will be unable to recover all their DSM expenses. In Wisconsin, the opposite was true—that is, the Commission was concerned that utilities would recover more than they spent on DSM because the future test year established an anticipated budget level and based rates on that level. If the utility spent less than this amount, there would be no adjustment. The utility would simply retain the difference between the actual and budgeted amount as a contribution to earnings. As a result, the Wisconsin PSC established the "conservation escrow account," sometimes called a balancing account. This mechanism entitles the utility to collect approved DSM expenditures from its ratepayers on a dollar-for-dollar basis. Therefore, if the utility underspends its DSM budget during the year, in the next rate case a credit equal to the amount of underspending is subtracted from the revenue requirement. If, on the other hand, the utility spends more than it budgeted for DSM, the under-recovered amount is included in the revenue requirement in the next rate case. This eliminates any incentive for the utility to underspend on DSM and also prevents losses from overspending. The utility recovers no more and no less than its actual DSM program costs.

The other accounting provision for DSM spending is capitalization of relevant expenditures. Any DSM expenditures thought to produce long-term benefits to the utility system are capitalized rather than receiving expense treatment. These expenditures are usually investments in customer equipment, such as rebates, loans, or shared savings contracts. The Wisconsin PSC has determined that the costs of providing energy audits, informational brochures and administrative services should be expensed, as it is much more difficult to attribute tangible benefits to such expenditures. Capitalization of DSM expenditures provides no financial incentive per se for the utility to invest in demand-side resources. To provide an incentive, the utility would have to expect to earn more than its required return on those investments.

Any investment that receives less than the required return

decreases the market value of the company.[1] Any investment that exactly earns the required return neither contributes to, nor detracts from, market value. Only if DSM investments earn more than the required return will the value of the company increase as DSM expenditures are increased. In that case the utility would have an incentive to increase DSM spending. The Wisconsin PSC's generic procedure for capitalizing DSM expenditures, however, does not call for a higher return for demand-side investments than for supply-side investments. The Wisconsin PSC has experimented with return on equity bonuses in some cases, but those were the exceptions rather than the rule.

The Wisconsin PSC capitalizes DSM expenditures for equity or fairness, not for financial reasons. If an asset produces system benefits over several years, it is unfair to charge current ratepayers the entire cost of the asset. This thinking is consistent with the treatment of supply-side investments, i.e., a power plant is depreciated over its useful life, thereby amortizing its cost over many years.

All these procedures—annual rate relief, future test years, and appropriate accounting mechanisms—provide for comprehensive and fair treatment of DSM expenditures and their impacts. For the most part, all of these procedures could be adopted by other states to help reduce some of the negative attitudes utilities may have about DSM. But such procedures alone may not be sufficient to create a positive environment for utility DSM spending. At least as important as the procedures is a regulatory philosophy that encourages the utility to invest in cost-effective demand-side resources.

Wisconsin's Regulatory Policies

Financial Integrity

The Wisconsin PSC places a premium on maintaining utilities' financial integrity. As a result, Wisconsin utilities have established strong financial track records and low business and financial risk profiles. For example, throughout most of the 1980s, the major Wisconsin electric utilities had the highest bond ratings and among the lowest stock betas of all the major U.S. electric utilities.[2] These high financial marks

---

[1] Capitalization of assets increases the accounting or book value of a company. But, according to financial theory, unless an asset earns its required return, capitalization will decrease the true or market value of the company.

[2] Beta is a measure of the systematic risk of a stock. The more a stock moves in tandem with the general stock market, the higher the beta. A stock with a low beta tends to move independently of the stock market in general. Low beta stocks can provide portfolio diversification that is not available from high beta stocks.

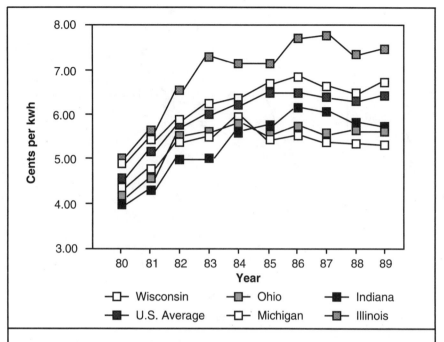

Figure 11-1. Average Rates Comparison: Wisconsin Versus Midwest States

were achieved with relatively high returns on equity and especially high common equity ratios.[3]

While some would claim that such financial prosperity comes at ratepayers' expense, the evidence does not support this claim. Wisconsin electric bills and rates are relatively low according to national standards. In a recent survey, the average costs per kilowatt-hour of the major Wisconsin electric utilities were among the lowest third of all utilities in the nation. Figures 11-1 through 11-4 compare Wisconsin's average rates and relative rate increases to those of other midwestern states and states with similar fuel mixes. These data confirm that Wisconsin electric utilities have lower rates and that the rate of *increase* since 1980 has been lower in Wisconsin than in any other state examined (Edison Electric Institute, 1980–1990).

---

[3] In the early 1980s, the Wisconsin PSC's allowed returns on equity were not high by national standards. In the mid to late 1980s, however, Wisconsin's allowed returns tended to be high, especially considering the low risk of the Wisconsin utilities.

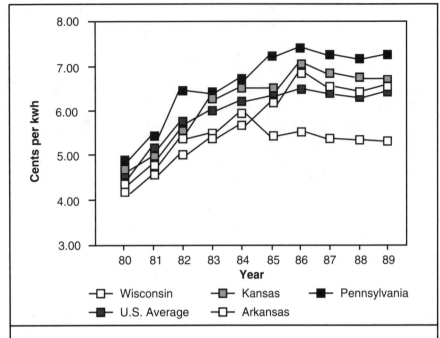

Figure 11-2. Average Rates Comparison: Wisconsin Versus Similar Fuel Mix States

In addition to relatively low rates, Wisconsin utilities offer customers opportunities to participate in comprehensive programs that allow them to further lower their bills by reducing their use of electricity. Wisconsin ratepayers in general apparently have no cause to be unhappy with the quality or cost of electric service.

The Wisconsin PSC believes that the financial integrity of the utility is crucial to meeting ratepayer needs. The evidence suggests that, over the long run, operational decisions (e.g., deciding whether or not to build a power plant or promote DSM) are much more important in determining utility bills and rates than are financial decisions (e.g., return on equity and capital structure). Keeping the utility financially strong increases its flexibility to make appropriate operational changes when necessary to meet changing circumstances (Kihm 1988).

While this focus on financial integrity may seem to be a boon to the utility and its stockholders, it has some strings attached. If a utility does not implement Wisconsin PSC policies in good faith, the financial decisions become less favorable. In some cases Wisconsin utilities

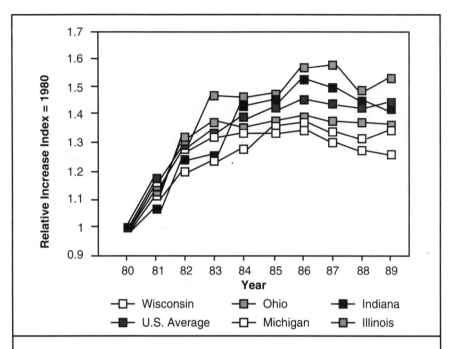

Figure 11-3. Relative Rate Increase Comparison: Wisconsin Versus Midwest States

have received return on equity penalties for failing to carry out PSC orders. Some have also received return on equity bonuses for good performance in areas other than DSM.[4] More importantly, the generally positive financial environment means the loss of some control by the utility. In Wisconsin, important operational decisions are often made by the PSC, not the utility.

Strategic Decisions

In Wisconsin, the following decisions are made by the PSC: whether a power plant should or should not be built, what type of power plant should be built; how much money should be spent on DSM programs annually; what transmission lines need to be upgraded; what the utili-

---

[4] In an order for WEPCO, for example, the Wisconsin PSC added 10 basis points to the utility's authorized rate of return because the utility took advantage of higher-than-expected revenues and invested in additional maintenance of its facilities, thus lowering future costs to ratepayers.

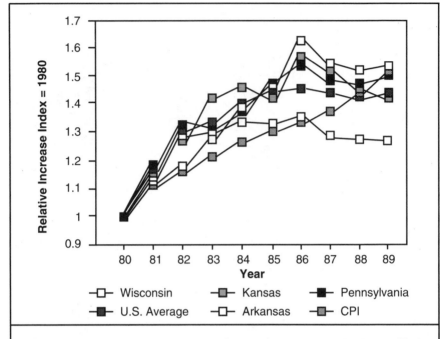

Figure 11-4. Relative Rate Increase Comparison: Wisconsin Versus Similar Fuel Mix States

ty's common equity ratio should be over the next few years; and what dividend policy is appropriate given the utility's circumstances. While the PSC's pro-active decision-making process requires substantial input from the utilities, as well as from commission staff, environmental and consumer intervenors, and other state agencies, the PSC, not the utility, makes the final decisions.

We believe that state commissions should make such public policy decisions. Although many utilities do consider the needs of other groups when making decisions, the utilities must heavily weigh the impacts of any decision on their companies and their stockholders. To expect the utility to integrate its needs with the views of other stakeholders is poor public policy. On the other hand, both the Wisconsin PSC and the utilities have a common goal: meeting customers' needs with the highest-quality utility services at reasonable cost.

Some major decisions have been vigorously opposed by the utilities. For example, in the late 1970s, the Wisconsin PSC denied utilities' requests to build several large nuclear power plants, ordering them to build smaller coal-fired plants instead. While these decisions turned out to greatly benefit stockholders and ratepayers, they were not easy decisions for the Wisconsin PSC to make at the time.

Specifically with respect to strategic DSM-related decisions, the Wisconsin PSC determines how demand-side resources should fit into the utility's integrated resource mix and the appropriate DSM spending levels. In some cases, the PSC has suggested much higher spending levels for DSM than those proposed by the utilities. The Wisconsin PSC staff is also involved with the details of DSM programs. The staff works with the utilities and interested intervenors to develop net-benefits goals for each market sector, establishes qualitative factors to be used along with the goals to judge utility performance, and approves marketing approaches, efficiency levels and rebate levels. While some state commissions might require additional staff to implement these activities, this type of regulation could certainly be replicated in other states.

In sum, the regulatory climate in Wisconsin is much different compared to regulation in many other states. The Wisconsin PSC is actively involved in strategic decision making and it emphasizes utility financial integrity. It has procedures that allow for full recovery of DSM-related costs and recovery of most, if not all, of the lost revenues from DSM programs. With this background on the Wisconsin environment, one could believe that Wisconsin would have no reason to adopt a DSM incentive mechanism. On the contrary, the PSC has not ignored the use of DSM incentive mechanisms. In fact, Wisconsin adopted several different DSM incentive mechanisms *prior* to the current national incentives movement.

Wisconsin's Experience with DSM Incentives

The Wisconsin Electric Power Company Incentive

The first DSM incentive mechanism used in Wisconsin was developed for Wisconsin Electric Power Company (WEPCO), which was ordered by the Wisconsin PSC to develop a massive conservation effort beginning in 1987 (Wisconsin PSC 1986). The order required all direct investments in conservation to be capitalized and allowed to earn the utility's current return. The incentive authorized by the Wisconsin PSC allowed WEPCO to earn an additional 1% return on its conservation investments for each 125 megawatts of demand savings it could achieve through its programs.

By April 1989, the utility reported that it had saved its first 125 megawatts and was entitled to begin receiving its performance incentive. The Wisconsin PSC accepted the utility's claim, but required that an audit be performed to review the utility's justification for the claimed savings. A consultant conducted an audit of the savings. This review did not examine actual bill savings, because the incentive was based on engineering calculations. Instead, the audit focused on the reasonableness of the engineering calculations of savings and the process used to implement the programs.

The audit indicated that a deduction of 16.6 megawatts from the utility's original estimate of 129.3 megawatts was warranted (Nichols et al. 1990). WEPCO protested the results, noting that the PSC's original order specified that "savings which might otherwise have been achieved without such programs (commonly referred to as windfall) will not be considered in these calculations." WEPCO claimed that to make the adjustments suggested by the audit "would change the rules of the game after the game has started." In ruling on this issue, the Wisconsin PSC agreed in part with WEPCO's argument (Wisconsin PSC 1990). The PSC did not make the suggested adjustments to WEPCO's numbers for the first program year, but it adopted the adjustments for subsequent years.

In dealing with the performance incentive issue, however, the Wisconsin PSC staff also noted some potential problems with WEPCO's program proposals. First, WEPCO proposed some programs that, if implemented, would have likely encouraged free ridership. Since the incentive mechanism did not include a reduction of savings to account for free-riders, the utility could profit from programs with high free-rider levels. These programs would install measures with no significant reductions in kilowatt-hour sales. Second, WEPCO also proposed to increase the estimates of energy savings for measures installed, based on new information from a revised residential end-use forecast. WEPCO intended to use these revised estimates in future claims of achievement of performance incentive goals. Since the savings levels were not fixed in advance, the incentive mechanism also gave the utility the impetus to claim higher per-unit savings. These examples highlight some of the problems associated with implementing a performance incentive.

WEPCO's New Proposal for an Incentive Mechanism

In its 1991 test year rate case, WEPCO proposed a new incentive mechanism that would allow shareholders to receive 10% of the net benefits achieved by WEPCO's DSM programs in 1991. The utility estimated the amount at slightly more than $5,000,000. WEPCO indicated that the incentive was needed "to maintain high performance levels in the area of least-cost planning" (WEPCO 1990a). The Wisconsin PSC staff opposed the utility's proposed incentive, saying "the desired behavior the performance incentive is supposed to encourage will occur without the incentive" (Wisconsin PSC Staff 1990).

In its order, the Wisconsin PSC denied WEPCO's proposed incentive mechanism, but found:

> In order to permit the company to accelerate the change in its corporate culture to an integrated least-cost planned entity by committing all its personnel to promote demand-side programs, it is reasonable to include dollars in the test year escrow account for demand-side performance incentives to *employees*. A reasonable amount for this purpose is $500,000. Employees . . . who are instrumental in achieving demand-side benefits *which exceed the normal expectations* should be eligible for incentives, whether or not their normal duties include implementation of demand-side programs" (Wisconsin PSC 1991; emphasis added).

Now, at least in WEPCO's case, managers and employees will be rewarded for reducing, rather than increasing, sales growth.

The Madison Gas and Electric Competition Incentive

In developing an incentive mechanism for Madison Gas and Electric Company (MG&E), the Wisconsin PSC staff believed it was appropriate to concentrate more on the psychological aspects of an incentive, rather than its financial impact. The staff noted a significant preoccupation on the part of Wisconsin utilities with *competition*, although opportunities for true competition among utilities are extremely limited. The staff asked, If the perception of competition stimulated this much activity and interest from the utilities, why not introduce competition into the demand-side marketplace?

In June 1988, the Wisconsin PSC directed MG&E to participate in a competition to provide demand-side services to its customers in three market sectors: multi-family rental; small commercial and industrial; and large commercial and industrial (Wisconsin PSC 1988). Three contractors, one for each sector, were chosen to provide DSM services to MG&E's customers. At the same time, MG&E was required to offer its own services to customers in all three sectors. Each group was allocated the same amount of funding. The objective was to see which organization could achieve the most savings with its available resources. While MG&E staff felt disadvantaged because the utility had to compete in all three sectors, it also had the advantage of being able to shift resources from one sector to another during the competition.

The complete details of the working of the competition pilot can be found elsewhere (DeForest and Berkowitz 1990; Vine et al. 1990). The "winners" and "losers" were determined by a scoring. To the "winner" went a small financial incentive based on the margin of "victory." The outcome of the competition was that the utility "won" in two of the sectors and received an incentive of about $200,000. The third sector was "won" by one of the competitors, which received about $40,000.

Several important effects were observed. First, and most important, the competition pilot *did* stimulate a significant increase in DSM program activity. MG&E's customers were offered substantially more DSM services and realized significant savings on their utility bills. Therefore, the customers were the true winners.

Second, as the competition progressed, it quickly became apparent that the potential for monetary reward was not as powerful a motivator as "winning" or "losing." Third, each of the competitors, both utility and non-utility, spent its available funds on cost-effective demand-side measures and achieved significant savings. Finally, a great deal of information was gathered to aid in future program development and regulatory approaches to DSM.

A side benefit was that other Wisconsin utilities, recognizing that they could be ordered into a similar competition, improved their programs without waiting for Commission action. Thus, the competition pilot ultimately provided an incentive even for utilities that were not involved.

Other Wisconsin Incentive Mechanisms

At the same time, the Wisconsin PSC also attempted to implement two other incentive mechanisms for Wisconsin Power and Light Company (WP&L) and Wisconsin Public Service Corporation (WPSC). The details of these incentives have been discussed elsewhere (Newman and Schoengold 1990). Briefly, WP&L was given an incentive as part of its shared-savings program. Under this program, WP&L would finance energy conservation measures for customers, who would repay the cost through the electric bill savings. The financing period would usually be four to five years, and the interest rate charged would normally be set at the utility's authorized rate of return. WP&L's incentive was the opportunity to earn higher interest on each contract, provided it could achieve a minimum of 10% bill savings. The greater the projected bill savings, the higher the interest it could charge.

WPSC's mechanism involved a modest goal of saving 10 million kilowatt-hours. The utility would earn an incentive for performance above the goal or be penalized for achievement below the goal. The maximum incentive was limited, however, to a range of minus $30,000 to plus $190,000. The utility's annual revenue requirement was $540 million.

The main difficulty with the WP&L and WPSC incentives was that the dollar amounts were insufficient to stimulate any significant action by the particular utility. They also suffered from emphasizing only energy savings (in one case excluding gas savings), and measuring savings was an issue, as it is with all performance incentives. WP&L had problems getting customers to participate at *any* rate of return. Adding an incentive for the utility just made the shared-savings offer less attractive to an already reluctant customer. Both the WP&L and WPSC incentive mechanisms failed miserably; the financial and the psychological effects of these mechanisms were insufficient to motivate the utilities to aggressively pursue DSM.

Policy Directions

Two of Wisconsin's incentive experiments—those involving WEPCO and MG&E—had some success, although the WEPCO program also suggested certain weaknesses of the approach. The WP&L and WPSC programs clearly failed, which led the PSC staff to ask whether *any* incentives were really necessary, beyond those that existed under current Wisconsin regulation.

Do Utility Stockholders Need Incentives?

Sales Growth and Stockholder Returns

It is commonly believed that since demand-side measures reduce utility sales, promoting demand-side measures is harmful to utility stockholders (Moskovitz 1989). This assumption is based on the premise that greater sales growth leads to higher stockholder returns. The Wisconsin PSC has rejected this assumption (Kihm 1991).

According to finance theory, growth and stockholder wealth are not necessarily related. Stockholder wealth is increased only if a project can deliver positive net present value. To produce positive net present value, a project must earn more than its required return. If an investment, no matter how large, earns exactly the required return, stockholder wealth is unchanged. If it earns less than the required return, the greater the investment, the more stockholder wealth declines.

Consider the likely impact of utility sales growth on investors. As

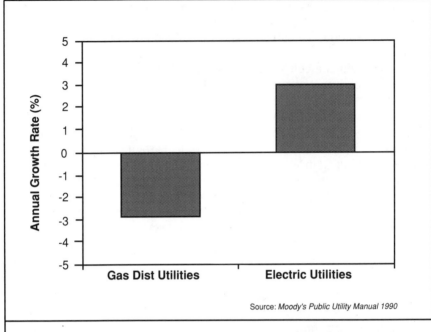

Figure 11-5. Annual Sales Growth for Utilities in Moody's Stock Indices

sales increase, the need for more plant capacity must eventually increase. To build new plant, the utility invests capital into the firm. Does this increase in capital investment increase stockholder wealth? According to finance theory, the answer depends on whether the earned return on investment is greater than the required return. But if regulation works correctly, utilities should, over the long run, only earn approximately the required return on their investments. Of course, if the utility is able to earn returns in excess of its required return, growth would increase the investor's wealth. On the other hand, if earned returns were less than the required return, growth would decrease the investor's wealth. Has growth led to greater returns for utility investors?

The answer is no. If anything, growth has led to less wealth for utility investors than a lack of growth. Figure 11-5 shows annual sales growth rates from 1972 through 1988 for utilities in the Moody's Gas Distribution Utility and Electric Utility stock indices.[5] While electric

---

[5] These are growth rates in sales as measured in kWhs and Btus, not dollars.

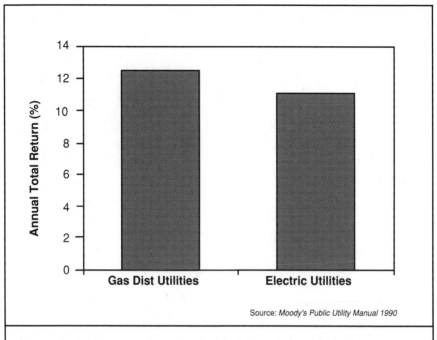

Source: *Moody's Public Utility Manual 1990*

Figure 11-6. Annual Returns to Stockholders of Moody's Utilities

sales increased during this period, natural gas sales decreased.[6] If growth is good for utility investors, electric utility stocks should have outperformed gas distribution utility stocks during this period.

Figure 11-6, which graphs total return to stockholders during the same period, shows that gas utility stocks, not electric utility stocks, were better investments. In other words, investing in the shrinking natural gas industry was more profitable than investing in the growing electric utility industry.

The strong relative performance of the gas distribution stocks was not due to the poor performance of electric utility stocks. In fact, the period from 1972 through 1988 produced extremely high real stockholder returns. Electric utility stocks outperformed the S&P 500 stock index returns over the 16-year period. Yet, despite the strong showing

---

[6] This is largely due to the fact that the real price of natural gas increased dramatically over this period, while the real price of electricity was approximately the same from year to year.

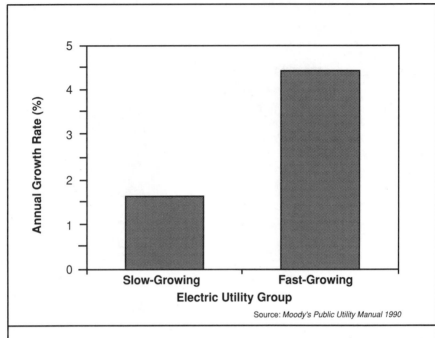

Source: *Moody's Public Utility Manual 1990*

Figure 11-7. Annual Sales Growth for Slow- and Fast-Growing Electric Utilities

of the electric utility stocks, the shrinking gas distribution companies' stocks produced even higher investor returns.

This high growth/low return relationship exists within utility industries as well as between them. For example, Figure 11-7 shows sales growth rates for fast-growing and slow-growing electric utilities.[7] Again, if growth is good for the stockholder, investing in the fast-growing group should have produced higher investor returns than investing in the slow-growing group.

Figure 11-8 shows that the faster the sales growth rate, the lower the investor return. The conclusion is clear. Simply because a utility increases its sales growth does not mean that its investors benefit. In fact, the investors are likely to be worse off in rapid-growth environments than in slow-growth or even negative-growth situations.

---

[7] The fast-growing group is the upper quartile in terms of sales growth for the Moody's Electric Utilities over the 1972–1989 period. The slow-growing group is the lower quartile of companies.

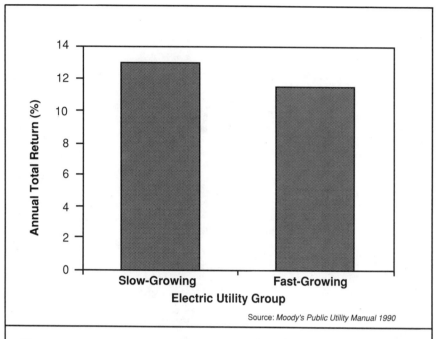

Source: *Moody's Public Utility Manual 1990*

Figure 11-8. Annual Total Returns to Stockholders of Electric Utilities

Could the negative implications of growth be a recent anomaly, one attributable to the overbuilding and nuclear construction programs in the late 1970s and early 1980s? That does not appear to be true. Figure 11-9 shows that even in the 1950s and 1960s, high-construction environments harmed utility investors, while low-construction conditions were prosperous for investors.

Wall Street's Perspective on Growth

The evidence presented in the preceding section is not new to the financial community. It is easy to find statements from Wall Street analysts suggesting that growth is not necessarily good for utility investors. Three examples are especially relevant.

First, a Goldman Sachs portfolio manager writing in *Financial Analysts Journal* found that from 1967 through 1987,

> . . . outperforming utilities [utility stocks with high investor returns] have had lower historical growth rates than their underperforming counterparts [utility stocks with low investor returns] . . . (Jones 1990).

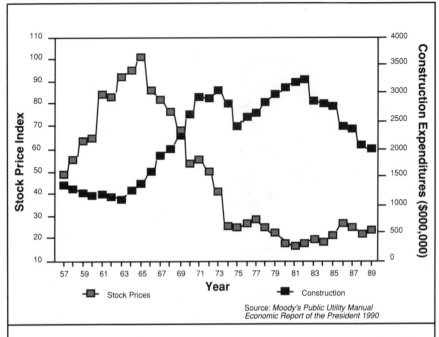

Figure 11-9. Stock Prices Versus Construction Expenditures for Moody's Electric Utilities

Another example of the "growth isn't good" position appears in *Value Line Investment Survey*'s recent comments on Delmarva Power's stock:

> . . . the stock has less interest as a total return vehicle. As shown in the operating statistics . . . last year's peak load was about 89% of Delmarva's generating capacity. That ratio is on the high side; most utilities strive for a ratio of 80% or less to ensure adequate supplies at peak periods. At the same time, load growth has been expanding at a good clip, reflecting the economic health of the service area. That means that capital outlays will have to remain high to finance the needed capacity increases. A fair amount of this capital will have to be obtained from public debt and equity offerings. The cost of this new capital may put a damper on share earnings and dividend growth at least through the 1993–95 period (Schlien 1990).

Finally, a leading credit analyst takes a similarly negative view of utility growth:

> Although above-average growth is viewed positively in an industrial company, it may be viewed negatively with respect to an electric utility.

An electric utility with above-average growth must necessarily have a high construction budget (Howe 1991).

These quotes are consistent with our findings: growth is not viewed as a positive indicator by utility investors. If reducing sales growth would not hurt utility investors, and probably would help, why then do we find utility managers demanding stockholder incentives for promoting demand-side measures? Do they perceive DSM as more risky than curtailing growth that occurs for other reasons?

Utility Managers' Aversion to Demand-Side Management: Perceived Risk and Corporate Culture

DSM and Risk. There is a perception among utility managers that demand-side management is more risky than supply-side investments; however, there is strong opposition to this view. For example, Hirst says that demand-side management offers considerable risk-reducing and flexibility benefits compared to even the lowest-risk supply-side measures (Hirst 1990). The Vermont Public Service Board has found that demand-side resources offer "comparative risk and flexibility advantages" compared to supply-side resources (Vermont PSB 1990). The Wisconsin PSC concluded that demand-side programs produce less financial uncertainty and less business risk (Wisconsin PSC 1989). Utilities that have aggressively promoted demand-side measures have found that demand-side measures are less risky than supply-side investments (NEES 1989). Demand-side investment is also politically more acceptable because customers receive direct benefits from reductions in their bills and there is less "rate shock" because DSM is added gradually. Finally, implementing DSM measures before constructing expensive generating facilities may lessen the potential "death spiral" of higher costs, higher rates, and ever-diminishing sales over which to spread fixed costs. Implementing DSM before supply-side measures can eliminate some of the elastic demand that might contribute to a "death spiral."

Many utility managers confuse "unfamiliar" with "risky." While it is true that aggressively promoting demand-side management with rebates and other marketing approaches is a new strategy for utilities, that does not mean it is a more risky strategy. Demand-side management offers a risk-reducing portfolio approach to meeting energy service needs and should be a welcome complement to total reliance on traditional supply-side resources.

Corporate Culture and DSM. We have seen that reducing growth need not decrease investor returns nor increase risk. Yet, it is difficult to find a utility corporate culture that supports reducing sales growth. Promoting sales growth has long been an objective for energy utilities.

American culture worships growth as an end in itself. For example, a major utility executive said that asking utilities to shrink their business opportunities is "un-American" (Rowe 1990). Economists have noted that psychological factors, namely power and prestige, are more closely correlated with sales volume than with profits or rates of return (Scherer 1980), with the result that utility managers tend to feel better when their company is growing.

Growth not only makes utility managers feel better, it makes them wealthier. While stockholder wealth is not related to company size and sales growth, managers' salaries tend to be (Marder 1990). But linking pay to size or growth is a poor way to motivate executives in any industry. According to an executive compensation expert,

> The relatively high association between firm size and executive compensation can only further fuel managements' natural inclination to grow businesses as fast as possible. There is no economic virtue in growth per se. "Bigger" does not automatically lead to "better" (Rappaport 1983).

Unfortunately, because of current compensation schemes, "bigger" does lead to more wealth for utility executives.

If sales growth is positively related to managerial salaries, but not positively related to investment returns, utility managers' pay must not be related to stockholder wealth accumulation. In fact, according to a recent study, electric utilities have the worst relationship between managerial pay and stockholder returns of any industry in this country (Jensen and Murphy 1990). The reason is that electric utility managers in America are not paid to increase stockholder wealth; they are paid to increase sales.

The Science of Positive Reinforcement

Before leaping into using stockholder incentives, state regulatory commissions should consider why many people think financial incentives will be successful in promoting DSM. Commissions should also examine whether incentives actually work the way people think they do.

Why are rate-of-return bonuses or shared-savings mechanisms believed to be necessary to get utilities to invest in DSM? One theory is that utilities do not make money from DSM, but only from investments in supply options. This view is commonly held since utility earnings are a function of the size of the rate base. Properly regulated utilities, however, are simply allowed to recover their costs plus a reasonable return on investment. Utilities do not make more money by increasing their investment in supply options; the cost of supplying energy simply goes up. On the other hand, in those jurisdictions where the commission does not allow a reasonable rate of return for utility investments, utilities will be reluctant to invest in *either* supply-side or demand-side options.

Another theory is that under traditional regulatory precepts utilities need to grow to satisfy investors. So, to get utilities to adopt policies that limit growth, it is necessary to provide bonuses for investors to make up for the loss of the benefits of unrestrained growth. But again, as we have previously shown, investors are not rewarded by growth. In fact, they are more likely to be rewarded by slow growth than rapid growth.

A third theory is that between rate cases, utility investors benefit from the increased revenues from sales greater than forecast and are hurt by reduced sales. While this would be true if there were no need to add new system capacity to meet increased loads, this is rarely the situation. As we have previously suggested, investors recognize the connection between sales growth and the need for system expansion and increased capital outlays. Through their actions in the stock market, investors have indicated that, given the choice, they would rather not have unrestrained growth along with the need for new capital investment. Financial analysts and investment advisors have shown the same preferences through their high ratings for low-growth utilities and low ratings for high-growth utilities. While the situation might be different for utilities that have substantial excess capacity and are experiencing no or negative growth, such utilities are rare. Policies should not be adopted based on these exceptions to the rule.

We have shown that the conventional reasons to justify the need for financial incentives do not hold up. Are there reasons that hold up? As discussed in the previous section, the managerial incentive to promote growth is strong, since managers tend to be rewarded for running larger companies, a fact that cannot be ignored. Financial incentives can help counter the managerial incentives against DSM, since managers like to report financial bonuses in annual reports to stockholders. It must be recognized, however, that stockholder incentives work not because they directly reward investors, but because they enable managers to feel good about operating successful DSM programs.

One could argue that if the result is successful, it is unimportant why stockholder incentives work; however, ignoring the actual workings of incentives can be counterproductive.

Utility investors are generally looking for safe investments that provide steady, reliable returns. They do not tend to look for spectacular but speculative returns. Utility investors analyze the expected returns from stocks and adjust the price they are willing to pay for a reasonable return for low risk. If the regulatory commission regularly allows rate-of-return bonuses, investors will simply adjust their expectations to take the bonuses into account and adjust the price accordingly. Then the regulatory commission will be stuck. Investors will view the removal or reduction of bonuses as the loss of an expected return and will complain and avoid the investment.

This is similar to a well-known effect in behavioral psychology: positive reinforcement for a particular behavior pattern can lead quickly to learning the rewarded behavior. It has also been demonstrated that when behavior is learned as a result of *regular* reinforcement (with rewards being given for every correct performance), such behavior is not learned very strongly. All it takes is a small number of withholdings of the reward for the learned behavior to be unlearned (Gleitman 1986).

Supporters of DSM incentives need to recognize they may be creating a situation in which utilities will implement DSM only if they get a bonus, and will drop DSM if the bonus is reduced or eliminated. If there is no other way to get utilities to implement DSM, perhaps we can accept the idea of paying higher returns (bribes) to get them to do it. But at the same time, we should ask why regulated, franchised, public utilities have to be bribed to do their jobs. If they are unwilling to do what society (as represented by regulatory agencies) expects of them for a "fair" rate of return, maybe someone else should be brought in to do the job. There are surely independent DSM operators who would be happy to develop and run DSM programs funded with ratepayer money, as the Madison Gas & Electric competition pilot vividly demonstrated.

Higher returns—bribes—are not necessary on a long-term basis to get utilities to implement DSM programs. Incentives can be useful as a start-up tool to get the attention of utility managers and induce a change in the corporate culture to promote efficiency rather than growth. *Occasional* financial rewards in recognition of *particularly* good performance can help to keep managers' attention on the goals of DSM.

Advantages of Wisconsin's Regulatory Approach

Wisconsin's regulatory approach permits the commission to closely control the energy resource future. We believe this approach, which is grounded in traditional regulatory principles, is good. One reason regulation developed in the first place was because society viewed the provision of utility services as too important to leave to the vagaries of the "free" market. We believe this is still true.

Making the traditional approach work for DSM requires certain conditions and a mindset in the regulatory commission and among its staff. Without these conditions and mindset, it will be difficult to induce utilities to expand DSM. We believe, however, it is possible for other states to adopt many of the procedures used in Wisconsin, if not the regulatory philosophy as well.

An important condition is a regulatory framework under which the commission has (or can create) the responsibility to authorize utility supply- and demand-side actions and marketing programs. While this authority is fairly common to commissions, it is not universal. If a commission does not have the right of prior authorization, but can only accept or reject utility actions after the fact, it may be difficult to institute some of the regulatory approaches we advocate. Further, under an after-the-fact regulatory scheme, it will be difficult for the commission to have any significant impact on DSM, no matter what incentive scheme is used.

It is useful, though not necessary, for the regulatory commission to use a future test year in rate cases. With a future test year, the commission can include the impacts of projected DSM programs in the sales forecast and revenue requirements. Future test years minimize the need for lost-revenue adjustments, either before or after the fact. The revenues from "lost sales" will be factored into the test year projections and so will not be lost.

Rate cases with reasonable frequency are also useful. If opportunities to adjust the sales forecast are infrequent, a revenue adjustment may be necessary.

For utilities to move strongly forward with DSM programs, assurance of full recovery of reasonable DSM expenditures is absolutely necessary. Programs should be reviewed ahead of time by the commission or its staff to determine reasonableness. This means that, on occasion, the commission may allow recovery of DSM costs that ultimately are found to exceed the benefits of their corresponding DSM savings. The alternative is a utility that will refuse to innovate and will use only the most proven and overly conservative approaches—resulting in underinvestment in DSM.

The Wisconsin PSC's approach uses the commission staff to review utility plans and advise the commissioners. It assumes a great deal of proactive involvement by the commission and its staff in planning and implementing DSM programs. If a state's regulatory framework does not allow the staff to work closely with the commission, implementing this approach might be difficult.

An alleged benefit of DSM incentives is that they minimize the work for the regulatory commission and its staff: a well-designed incentive scheme is established that reduces the need for further commission encouragement and oversight of DSM. We believe that to make an incentive approach work requires as much effort as the more traditional approach we are recommending. With incentives, large sums of money may be riding on the success of the DSM programs; consequently, the utilities will use whatever means they can to make the results look good. The commission will need to devise ways to monitor program results to verify the utilities' claims of program success. This will require extensive monitoring by staff or consultants. If outside consultants are chosen, staff effort will still be needed to ensure the accuracy of the consultants' conclusions. Only through extensive review can the commission be assured that utility programs are making reasonable progress and that releasing incentive payments is justified.

Conclusion

Regulation of DSM programs should be accomplished much as regulation of other facets of the utility industry. As explained by George Sterzinger of the Vermont Department of Public Service (1990):

> Conservation programs must be developed within the context of existing regulatory policies, and must rely upon those checks and balances to assure that the programs are efficiently and fairly developed. This interest is not motivated by a fondness for the present regulatory structure in the abstract. Rather, it is motivated by a belief that any program which goes forward while ignoring the balances that are the essence of regulatory policy will produce consequences which will undermine broad public acceptance of conservation programs.

Wisconsin's experience suggests that before state commissions jump into the morass of providing utility stockholders with financial incentives for aggressive DSM programs, they would be wise to consider what can be accomplished through existing regulatory procedures and authority. Adopting exotic ratemaking mechanisms may do more harm than good toward achieving long-term DSM savings.

Procedures that should be adopted instead of DSM stockholder incentives include: full cost recovery for DSM expenditures; annual rate reviews with forward-looking test years; strong regulatory policies to ensure financial integrity of utilities; use of employee incentives for DSM and competition-type mechanisms to encourage good performance; and proactive commission involvement in key strategic decisions to ensure that ratepayers' interests are protected.

References

DeForest, Wayne, and Paul Berkowitz. 1990. "DSM Competition: A New Regulatory Strategy." In *Proceedings of the ACEEE 1990 Summer Study on Energy Efficiency in Buildings.* Washington, D.C.: American Council for an Energy-Efficient Economy.

Edison Electric Institute. 1980–1990. *Statistical Yearbook of the Electric Utility Industry,* various volumes. Washington, D.C.: Edison Electric Institute.

Gleitman, Henry. 1986. *Psychology.* New York: William Norton and Company.

Hirst, Eric. 1990. "Flexibility Benefits of Demand-Side Programs in Electric Utility Planning." *The Energy Journal* 11. January. P. 160.

Howe, Jane Tripp. 1991. "Credit Analysis for Corporate Bonds." In *The Handbook of Fixed Income Securities,* Frank J. Fabozzi, ed., p. 364. Homewood, Ill.: Business One Irwin.

Jensen, Michael and Kevin Murphy. 1990. "CEO Incentives: It's Not How Much You Pay, But How." *The Harvard Business Review* 68. May–June. Pp. 138–153.

Jones, Robert. 1990. "Designing Factor Models for Different Types of Stock: What's Good for the Goose Ain't Always Good for the Gander." *Financial Analysts Journal* 46. March–April. P. 28.

Kihm, Steven G. 1988. "Analysis of the Public Service Commission of Wisconsin's Energy Utility Capital Structure Policy." Madison, Wisc.: Wisconsin Public Service Commission.

Kihm, Steven G. 1991. "Why Utility Stockholders Don't Need Financial Incentives to Support Demand-Side Management." *The Electricity Journal* 4. June. Pp. 28–35.

Marder, David. 1990. "Executive Incentive Compensation." *The Electricity Journal* 3. December. Pp. 104–108.

Moskovitz, David. 1989. *Profits & Progress Through Least-Cost Planning.* Washington, D.C.: National Association of Regulatory Utility Commissioners.

New England Electric System. 1989. "Power by Design: A New Approach to Investing in Energy Efficiency." Westborough, Mass.: New England Electric System.

Newman, Paul and David Schoengold. 1990. "Back to Basics: An Innovative Approach Called Traditional Regulation." In *Proceedings of the Innovations in Pricing and Planning Conference, May 2–4, 1990.* Palo Alto, Calif.: Electric Power Research Institute.

Nichols, David, Gail Katz, Daljit Singh, Neil Talbot, and Elizabeth O.

Titus. 1990. *Savings from the SMART MONEY Program: An Audit of the 125 Megawatt Demand Reduction.* Boston, Mass.: Tellus Institute.

Rappaport, Alfred. 1983. "How to Design Value-Contributing Executive Incentives." *Journal of Business Strategy* 4. P. 50.

Rowe, John. 1990. "Making Conservation Pay: The NEES Experience." *The Electricity Journal* 3. December. P. 20.

Scherer, F. M. 1980. *Industrial Market Structure and Economic Performance.* New York: Houghton Mifflin.

Schlien, Milton. 1990. "Delmarva Power & Light." *Value Line Investment Survey* 46. September 21.

Sterzinger, George. 1990. "The Role of Utility Incentives in Designing Fair and Efficient Conservation Programs." Montpelier, Vt.

Vermont Public Service Board. 1990. "Least-Cost Investments, Energy Efficiency, and Management of Demand for Energy." *111 PUR 4th*, p. 422. June 22.

Vine, Edward, Odon De Buen, and Charles Goldman. 1990. *Stimulating Utilities to Promote Energy Efficiency: Process Evaluation of Madison Gas and Electric's Competition Pilot Program.* Berkeley, Calif.: Lawrence Berkeley Laboratory.

Wisconsin Electric Power Company. 1990a. Testimony of Dale A. Landgren, Docket No. 6630-UR-104. WEPCO 1991 Test Year Rate Case. Milwaukee, Wisc.

Wisconsin Electric Power Company. 1990b. Letter to the Wisconsin PSC from David K. Porter, dated June 8, 1990. Milwaukee, Wisc.

Wisconsin Public Service Commission. 1986. Order in Docket No. 6630-UR-100. WEPCO 1987 Test Year Rate Case. Madison, Wisc.

Wisconsin Public Service Commission. 1988. Order in Docket No. 3270-UR-102. MG&E 1988–1989 Test Year Rate Case. Madison, Wisc.

Wisconsin Public Service Commission. 1989. Order in Docket No. 05-EP-5. Advance Plan 5. Madison, Wisc.

Wisconsin Public Service Commission. 1990. Letter to Wisconsin Electric Power Company, dated October 25, 1990. Madison, Wisc.

Wisconsin Public Service Commission. 1991. Order in Docket No. 6630-UR-104. WEPCO 1991 Test Year Rate Case. Madison, Wisc.

Wisconsin Public Service Commission Staff. 1990. Rebuttal Testimony of Anita Sprenger Docket No. 6630-UR-104. WEPCO 1991 Test Year Rate Case. Madison, Wisc.

Chapter 12

Does the Rat Smell the Cheese? A Preliminary Evaluation of Financial Incentives Provided to Utilities

Steven M. Nadel and Jennifer A. Jordan

Introduction

In the late 1980s, the concept of providing financial incentives to utilities to pursue demand-side management (DSM) programs took the utility industry by storm. As of November 1991, incentive mechanisms were approved by utility commissions in 21 states (Chapter 2). These approvals were primarily based on the assumption that since utilities are private, profit-maximizing companies, if DSM is made more profitable, utilities will increase program activity levels. While this presumption makes intuitive sense, there has been no objective analysis on whether utilities do respond this way. This chapter attempts to fill this void by objectively and subjectively analyzing how utilities have responded to DSM incentives.

However, in undertaking such an analysis, considerable caution must be used. Incentives have been used only a few years; many have been in place less than a year. The data on post-incentive experience are limited and may change. Thus, our analysis is a preliminary evaluation—a complete evaluation cannot occur without additional data including more information about present incentive mechanisms as well as experience with mechanisms likely to be implemented in the next few years.

Approach

For this study we examined the DSM activity of 17 utilities with some DSM incentive mechanism approved as of January 1991. The 17 utilities include only those that receive positive financial incentives for DSM. Table 12-1 lists these utilities, and includes basic descriptive information about each utility and its incentive mechanism. However, incentives are not the only financial mechanism that affects utility interest in DSM; other important mechanisms include cost recovery for DSM expenses, recovery of lost revenues, decoupling of profits from sales (discussed in Chapters 3 and 4), ratebasing of DSM investments (discussed in Chapter 5), and utilities operating as an energy service company (discussed in Chapter 9). Since these other mechanisms do not provide a positive incentive to pursue DSM, they are not included in this analysis. However, many utilities that receive incentives also benefit from these other mechanisms, as shown in Table 12-1.

The 17 utilities in Table 12-1 represent utilities that began earning incentives from December 1986 to January 1991. Utilities in the state of Washington that receive a small bonus return on DSM investments since 1980 are not included (see Chapter 2), because, after 11 years, it is difficult to get appropriate data for the period immediately before and after incentive approval. However, a previous appraisal of these incentives found only modest impacts (Blackmon 1991).

In addition to examining data for the 17 utilities with DSM incentives, we examined data for 14 other utilities that do not receive incentives, but which border utilities that do. The latter group, which is listed in Table 12-2, was the control group for the analysis. Wherever possible, the control group utilities are in the same states, and regulated by the same commissions as utilities receiving incentives. However, in New York State and California, all investor-owned utilities receive incentives. In these cases, we used the neighboring states of New Jersey, Pennsylvania, Nevada, and Oregon for control utilities. While these utilities do not receive incentives, they do have cost recovery for DSM expenses, and one receives lost-revenue recovery (see Table 12-2).

In order to measure the impact of incentives on DSM activity, a three-pronged approach was used. First, we examined annual DSM savings (kWh and peak kW) and DSM expenditures for the years before and after incentives were approved. Second, we examined long-range DSM plans for the 1991–2000 period (or the nearest available dates), comparing the plans prepared immediately before and after incentives were approved. Third, we interviewed staff at each utility, each utility commission, and active intervenor groups.

Utility	Date Incentive Approved	Program Cost Recovery	Lost-Revenue Recovery	Decoupling	Post-Incentive Year Used in Analysis	Incentive Value for Post-Year Used in Analysis (1989 gross revenues)	Brief Description of Incentive
NEES							
Narragansett	12/27/89	yes	no[a]	no	1991	0.40%[b]	Shared savings
Granite State	8/7/90	yes	no[a]	no	1991	0.62%	Shared savings
Mass. Electric	3/3/90	yes	no[a]	no	1991	0.28%	Bounty
Western Mass. Elec.	6/29/90	yes	yes	no	1991	0.26%	Bounty
United Illuminating	1/24/90	yes	yes	no	1990	0.10%	Bonus ROR on DSM based on cost-to-savings ratio
Pacific Gas & Elec.	8/29/90	yes	yes	yes	1991	0.37%	Shared savings; markup on expenditures
San Diego G&E	9/88	yes	yes	yes	1989	0.26%	Bounty
	8/29/90	yes	yes	yes	1991	0.24%	Shared savings; markup on expenditures
So. California Ed.	8/29/90	yes	yes	yes	1991	0.09%	Ratebasing; markup on expenditures
Public Service of CO	11/28/90	yes	no	no	1991	n/a[c]	Shared savings
Portland Gen. Elec.	12/7/90	yes	yes	no	1991	0.04%	Shared savings
Wisconsin Elec.	12/30/86	yes	no	no	1987	0.03%[d]	Bonus ROR on DSM based on MW savings target

Consolidated Ed.	8/30/90	yes	no	1991	0.49%	Shared savings[e]
Orange & Rockland	9/12/89	yes	no	1990	0.12%	Shared savings
	8/30/90	yes	yes	1991	0.94%	Adjustment to overall ROR based on actual performance versus goals
LILCo	6/20/89	yes	no	1990	0.13%	Bonus ROR based on MW savings target; 120% of lost revenues
Central Hudson	8/31/90	yes	no	1991	0.13%	Shared savings
	1/4/90	yes	no	1990	0.18%[f]	Shared savings
Rochester Gas & Elec.	7/6/90	yes	no	1991	0.12%	Shared savings
NY State Elec. & Gas	1/4/90	yes	no	1990	0.13%[g]	Shared savings
Niagara Mohawk	12/15/89	yes	no[h]	1990	0.06%	Shared savings
Consumers Power	3/13/90	yes	no	1991	0.06%	Bounty[i]

a Lost-revenue recovery is less of an issue for NEES for the following two reasons: 2 of the 3 companies have a forward versus historical test year, and all 3 companies purchase power from a wholesale company. Nevertheless, NEES is currently evaluating whether or not they are incurring lost revenues due to DSM. On the wholesale level, NEES has a tracker allowing for adjustments in sales within a certain boundary; thus NEES has a decoupling mechanism of sorts. Incentive earnings for 1991 are mid-1991 estimates.

b All earnings are at the post-tax level.

c Public Service of CO has made no estimate of the incentive expected as a result of their 1991 activities.

d Earnings were calculated taking the average of the following: 1) fraction of MW savings from '87–'90 due to '87 programs multiplied by incentive expected for meeting 225 MW savings mark; 2) the fraction of '87–'90 DSM budget spent in '87 multiplied by incentive expected for meeting 225 MW mark.

e Con Ed has a new incentive as of 3/6/91 which is similar to the latest O&R incentive mechanism.

f The incentive earnings are for June 1990 through May 1991.

g The annualized incentive earnings were calculated by multiplying (earnings for June 1989 through December 1990)×(12/19).

h In a recent rate case Niagara Mohawk adopted an ERAM. However, it does not apply to DSM programs. The company is still calculating lost-revenue effects.

Utility	State	Neighboring Utilities with Incentives	Cost Recovery	Decoupling	Lost-Revenue Recovery
Table 12-2. Control Group Utilities					
Boston Edison	MA	NEES	yes	no	no
Commonwealth Electric	MA	NEES	yes	no	no
Connecticut Light & Power	CT	UI, WMECo, NEES	yes	no	no
Eastern Utilities Associates	MA/RI	NEES	yes	no	no
Jersey Central Power & Light	NJ	Con Ed, O&R	yes	no	no
Metropolitan Edison	PA		yes	no	no
Pennsylvania Power & Light	PA	NYSEG, O&R	yes	no	no
Public Service Electric & Gas	NJ	Con Ed, O&R	yes	no	no
Nevada Power	NV	So. Cal. Ed.	yes	no	no
Pacific Power & Light	CA/OR	PG&E, PGE	yes	no	no
Sierra Pacific	NV	PG&E, So. Cal. Ed.	yes	no	no
Detroit Edison	MI	Consumers Power	yes	no	no
CENTEL Electric	CO	PS of Colorado	yes	no	no
Wisconsin Power & Light	WI	WEPCo	yes	no	yes

For the analyses of recent and planned DSM activity, a series of ratios were calculated based on the level of post-incentive activity to pre-incentive activity for each utility. One set of ratios was based on the ratio of kWh savings, peak kW savings, and DSM expenditures in the first year following approval of incentives to the analogous value in the year preceding approval of incentives.[1] Another set of ratios was

---

[1] All calculations are based on calendar years. If the incentive was approved in January or February, the year of approval is taken as the post-incentive year. If the incentive was approved in November or December, the year of approval is taken as the pre-incentive year. If the incentive was approved in any other month, the preceding year is the pre-incentive year and the following year is the post-incentive year. Several utilities have used different incentive mechanisms at different times. For these utilities, the pre-incentive year precedes

based on the level of kWh savings, peak kW savings, and DSM expenditures included in each utility's post-incentive long-range plan to the analogous values in the utility's pre-incentive long-range plan. Long-range values were for the 1991–2000 period, or the closest approximation for which data could be obtained. In total, six ratios were calculated for each utility; three ratios based on data for recent years and three based on long-range plans. For each set of three ratios, one is based on kWh savings, one on peak kW savings, and one on DSM expenditures.

In calculating these ratios for each utility, several guidelines were used. First, we tried to ensure that data for each utility were calculated consistently, so that the numerator and denominator for each ratio are based on equivalent data. After allowing for this constraint, we standardized the definitions for each type of data as much as possible. Data on kWh and kW savings are generally net savings at the customer level, and are based on engineering estimates (see Chapter 10 for a discussion on these terms). Net savings were usually calculated by estimating the total savings achieved by program participants and then subtracting savings attributable to free riders. Savings figures are for the measures installed during the year for which credit is taken and are annualized. For load management programs, kW savings generally only include customers who first sign up in the year when credit is taken—re-enrollments of previous participants are usually not included. Demand (kW) savings generally coincide with each utility's peak demand. DSM expenditures are in 1990 dollars.

Results

This discussion of results is divided into two parts. First, the results of the data analysis for all the utilities examined are discussed. The emphasis in this section is on overall trends in the analysis, and not on results for specific utilities. Second, results for individual utilities are discussed, based on both the data analysis and the interviews. This discussion is grouped by region, beginning with New England and continuing through New York, the West Coast, Wisconsin, Colorado, and Michigan. Since some of the interviews were "off the record," we do not attribute comments to individuals, but refer to the general perspective of the commenter, such as utility or regulatory personnel.

---

all incentives, and the post-incentive year follows all incentives. For control utilities, the pre- and post-years used to calculate the ratios are based on the pre-incentive and post-incentive year of neighboring utilities. In some cases the post-incentive year was 1991. In these cases we generally combined data on actual results for the first part of the year with data on estimated results for the remaining part of the year.

Overall Analysis

Table 12-3 describes six different ratios of DSM activity for each utility. Utilities receiving incentives are listed at the top of the table; utilities without incentives (control utilities) are listed below. The number of observations varies from ratio to ratio depending on the amount of data obtained from each utility.

For each ratio and each group of utilities—those with and without incentives—the median ratio is reported. We chose median values because a few high values produce distorted average values, particularly for analyses such as this one with small sample sizes. Overall, for each ratio, and each group, the median utility reports substantially higher levels of actual and planned DSM activity in the post-incentive period than in the pre-incentive period. For the utilities with incentives, the median increase ranges from a 115% increase in planned MW savings from 1991 to 2000 to a 250% increase in kWh savings in the first year after incentive approval. For the utilities without incentives, the median increase ranges from 0 to 70%.

For all six ratios, the median increase in DSM activity among utilities with incentives is greater than the median increase among utilities without incentives. The median increase for the with-incentives group exceeds the median increase for the without-incentives group by 169% for recent-year kWh savings; 120% for recent-year kW savings; 43% for recent-year expenditures; 75% for planned GWh savings; 43% for planned MW savings; and 29% for planned expenditures. The impact of incentives on recent-year activity is apparently greater than the impact on long-range plans. The reason for this finding is unclear; utilities may be waiting to see if incentives are permanent or to see if DSM programs are truly a viable replacement for supply-side resources before fully committing to DSM.

To see if these differences are statistically significant, we conducted a statistical analysis using the rank-sum statistical test, which compared the ratios for the with-incentives and without-incentives groups. The rank-sum test is a nonparametric statistical test designed for samples, such as small samples, which do not necessarily follow the normal distribution (Hoel 1971). Results of this analysis are summarized at the bottom of Table 12-3. These results are reported in terms of significance levels, where the significance level is the probability that the difference between two groups is due to random chance, and not to some intrinsic difference between the two groups. Thus, a significance level of .05 represents a 5% probability that random chance would account for observed differences between two groups. Statisticians usually look for a significance level below .05 or .1 before

Table 12-3. Ratio of DSM Activity Between Post-Incentive and Pre-Incentive Periods

Utilities with Incentives:

Annual Data						Forecasted Data					
GWh Savings (value in post-inc. yr)/(value in pre-inc. yr)		MW Savings		DSM Expenses		GWh Savings (sum from post-inc. plan)/(sum from pre-inc. plan)		MW Savings		DSM Expenses	
LILCo	7.1	RG&E	10.3	WMECo	8.1	NMPC	15.1	LILCo	10.1	NMPC	11.5
SDG&E	6.2	ConEd	6.6	O&R	5.5	SDG&E	11.1	SCE	6.0	SDG&E	7.8
RG&E	6.0	PSCO	6.0	RG&E	4.4	NYSEG	8.6	PG&E	4.8	ConEd	6.0
ConEd	5.6	PGE	4.3	PSCO	3.5	SCE	4.9	CenHud	4.3	SCE	4.2
PGE	4.3	WMECo	3.1	ConEd	3.5	CenHud	4.3	WEPCo	3.7	LILCo	2.7
CenHud	3.5	UI	2.7	PGE	3.4	WEPCo	4.0	NYSEG	3.4	PG&E	2.3
SCE	3.4	O&R	2.2	SDG&E	3.3	UI	2.8	NMPC	2.6	WEMCo	2.1
WMECo	3.2	PG&E	2.2	NMPC	3.1	PG&E	2.8	ConEd	2.2	O&R	1.8
PG&E	2.3	SDG&E	1.6	ConsPwr	2.7	ConsPwr	2.1	WMECo	2.1	NEES	1.7
UI	1.8	SCE	1.0	PG&E	2.3	WMECo	2.0	ConsPwr	2.0	ConsPwr	1.3
NEES	1.0	NEES	0.9	CenHud	2.1	NEES	2.0	SDG&E	1.3	PGE	1.2
		NMPC	0.7	LILCo	2.1	O&R	1.9	NEES	1.2	UI	1.0
		LILCo	0.3	NEES	2.0	PGE	1.1	PSCO	1.2		
				SCE	1.9	PSCO	0.6	PGE	1.1		
				NYSEG	1.5			UI	0.9		
				UI	1.3			O&R	0.7		

Utilities without Incentives:

Sample 11 Median 3.5		Sample 13 Median 2.2		Sample 16 Median 3.0		Sample 14 Median 2.8		Sample 16 Median 2.15		Sample 12 Median 2.2	
SP	3.0	PP&L	4.0	ComElec	6.3	EUA	4.9	EUA	7.2	EUA	3.5
PP&L	2.4	SP	2.4	DetEd	3.8	SP	2.8	DetEd	6.0	BECo	2.2
CL&P	2.0	CL&P	1.9	EUA	3.3	WI P&L	2.8	NevPwr	2.4	SP	2.1
BECo	1.9	PSE&G	1.6	BECo	2.7	NevPwr	2.1	CL&P	1.7	DetEd	1.7
ComElec	1.9	NevPwr	1.4	PP&L	2.6	CL&P	1.6	SP	1.5	CL&P	1.6
EUA	1.3	EUA	1.0	CL&P	2.1	BECo	1.2	BECo	1.3	NevPwr	1.2
NevPwr	1.3	CENTEL	0.9	PA P&L	1.4	MetEd	1.2	PSE&G	1.0	PSE&G	1.2
PSE&G	1.0	PA P&L	0.6	NevPwr	1.3	PSE&G	1.0	WI P&L	0.8		
MetEd	0.5	BECo	0.6	PSE&G	1.1	DetEd	0.7	MetEd	0.3		
PA P&L	0.3	MetEd	0.6	MetEd	1.0						
JCP&L	0.2	JCP&L	0.2	JCP&L	0.6						
Sample	11	Sample	11	Sample	11	Sample	9	Sample	9	Sample	7
Median	1.3	Median	1.0	Median	2.1	Median	1.6	Median	1.5	Median	1.7
Signif	0.002	Signif	0.030	Signif	0.063	Signif	0.066	Signif	0.231	Signif	0.156

considering a result statistically significant. Using this criteria, differences between the two groups in annual kWh and MW savings are significant at the .05 level, and the differences in annual expenditures and planned GWh savings are significant at the .1 level. Differences in planned expenditures and MW savings do not pass the .1 significance criteria—these differences were only significant at the .24 and .16 levels respectively.

According to these findings, there are statistically significant differences in DSM activity between the with-incentives and without-incentives groups that could be due to the incentives, or they could be due to other factors linked with the availability of incentives. The issue of causality is explored later in this paper.

While the average utility with incentives increased DSM activity more than the average utility without incentives, there are several exceptions. For example, Detroit Edison, Eastern Utilities, and Sierra Pacific, all utilities without incentives, often had ratios higher than the median of the without-incentives group. Likewise, Consumers Power, New England Electric, Portland General Electric, Public Service of Colorado, and United Illuminating, all utilities with incentives, often had ratios lower than the median of the without-incentives group. Reasons for these outliers are explored later in this chapter.

In addition to comparing the with-incentives and without-incentives groups to determine if incentives, regardless of type, have an impact, we also compared ratios among the with-incentives group to see if patterns emerged. We wondered if specific incentive mechanisms produced higher or lower ratios, and if more generous incentives (higher earnings to the utility) resulted in higher ratios.

Comparing different incentive mechanisms produced no clear results, primarily because most of the incentives were shared savings, and sample sizes for the other mechanisms were too small to justify making conclusions.

For the analysis on the effect of incentive amount on levels of DSM activity, we compared the annual ratio for GWh savings with the incentive amount as a percent of gross revenues. This analysis, illustrated in Figure 12-1, also found no relationship.

Regional Analysis

New England. In New England, three utilities with incentives are included in the analysis—New England Electric, Western Massachusetts Electric, and United Illuminating. When comparing DSM activity to neighboring utilities without incentives, there is apparently no relationship between incentives and increased DSM activity. Figure

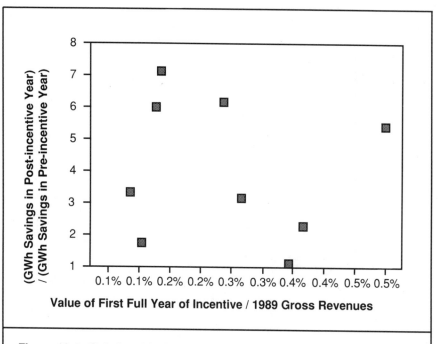

Figure 12-1. Relationship Between Level of DSM Activity and Amount of Incentive

12-2 plots recent annual GWh savings as a percent of GWh sales for New England utilities with and without incentives. The utilities without incentives are saving as much, or more, than the utilities with incentives.[2]

This appears to be true for several reasons. First, all the utilities included in the analysis have been involved in collaborative program design processes in which the Conservation Law Foundation and other interested non-utility parties worked with each utility to develop DSM programs. These efforts primarily took place during 1988 and 1989,

---

[2]   The results illustrated in Figure 12-2 must be interpreted with extreme caution because different utilities use different techniques to calculate energy savings. For example, the Commonwealth Electric and Eastern Utilities numbers (both utilities without incentives) are based entirely on engineering estimates and include savings achieved by free-riders. The numbers for New England Electric exclude savings by free-riders and are based in part on impact evaluation studies (differences between engineering estimates and impact evaluation results are discussed in Chapter 10).

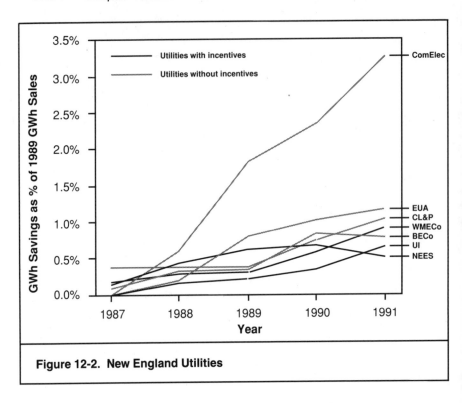

Figure 12-2. New England Utilities

and resulted in substantially increased DSM activity in all cases (Raab and Schweitzer 1992).

Second, the utility commissions in the states served by these utilities have been fairly aggressive in encouraging utilities to pursue DSM programs. For example, the Massachusetts commission in 1986 reduced Boston Edison's rate of return due to lack of progress on DSM (essentially a negative incentive), and in 1989, as part of a settlement associated with a nuclear plant outage, required Boston Edison stockholders to invest $75 million in DSM programs over a three-year period. Similarly, the Connecticut commission ordered Connecticut Light and Power to participate in the nation's first collaborative program design process.

Third, environmental issues in the densely populated regions served by these utilities make it difficult to site and license new power plants and transmission lines. Faced with power shortages in the late 1980s, New England utilities found that DSM was the least expensive, most politically expedient resource available.

Fourth, the desire to provide good customer service has led many of these utilities to offer extensive DSM programs, particularly at Commonwealth Electric. The Massachusetts commission ordered Commonwealth to raise rates to electric heat customers, producing enormous customer protests. Among the strategies Commonwealth used to address this dissatisfaction was expanded DSM programs to help consumers reduce electric bills.

Fifth, in the case of United Illuminating (UI), the incentive is rather small. UI calculates that it will be eight years before the utility profits from its incentive, and even then, the profit will be small.

Sixth, in the case of New England Electric and United Illuminating, demand-side management efforts *before* the incentive were among the more extensive in the nation, leaving less room for further growth.[3]

Seventh, savings estimates for two of the without-incentives utilities, Commonwealth Electric and Eastern Utilities Associates, may be inflated because these two utilities *include* free-riders in their savings estimates.

In this climate, the impact of incentives cannot generally be discerned from the data. However, discussions with the individual utilities and regulators involved indicate that incentives are having some impact. Perhaps the largest impact was at Western Massachusetts Electric (WMECo) where utility staff, regulators, and intervenors report an improved attitude towards DSM by senior management following award of the incentive. According to one observer, the attitude towards DSM at WMECo is much more positive than the attitude at Connecticut Light & Power, a sister-company that does not receive an incentive. Likewise, at New England Electric and United Illuminating, inside and outside observers report that incentive has reinforced management commitment to DSM. In particular, observers report that the incentives have quieted DSM skeptics in each utility.

Incentives have been in place in New England for more than a year, and utilities and regulators are beginning to make observations about the relative advantages and disadvantages of different incentive mechanisms. For example, New England Electric, which receives a bounty incentive ($/kW and kWh saved) in Massachusetts and a combination shared-savings and share-of-gross-benefits incentive in Rhode Island and New Hampshire (described in Chapter 6), reports that the bounty is simpler to administer. However, regulators claim that with the bounty, there is little incentive to control DSM costs, except to keep the program cost-effective. Regulators, intervenors and the utility agree that the shared-savings mechanism tends to encourage

---

[3] For example, of the 31 utilities included in this study, in 1989, NEES and UI had the third- and fourth-highest kWh savings from DSM programs as a percent of kWh sales.

programs for commercial and industrial customers, where net benefits of programs (value of savings minus program costs) are generally high, and to discourage residential programs where net benefits are generally low.

Incentives in Massachusetts are based on measured-savings results; therefore, extensive program evaluation activities are now underway. In its first evaluation report, Massachusetts Electric, a subsidiary of New England Electric, found that measured savings were often less than prior engineering estimates (Massachusetts Electric 1991), and as a result, the incentive is likely to be somewhat smaller than hoped for. Consequently, Massachusetts utilities are considering program modifications that may increase the actual savings.

New York State. In New York State, incentives have been approved for all seven investor-owned utilities. Most incentives were approved in 1990, although Long Island Lighting and Orange & Rockland's initial incentives were approved in 1989. Compared to utilities in the neighboring states of New Jersey and Pennsylvania, DSM activity in New York State has dramatically increased in the past two years (see Figure 12-3).[4] Based on discussions with New York utilities, regulators, and intervenors, a number of reasons account for this dramatic increase in DSM activity.

Perhaps the most important reason for the increase in activity, according to several utility respondents, is that the N.Y. Public Service Commission has ordered the utilities to increase DSM activity. However, according to many of the utilities, the availability of incentives and lost-revenue recovery mechanisms also influenced DSM activity. For example, one intervenor noted that in some cases the amount the utility received using the lost-revenue recovery mechanism was greater than the amount received using the incentive mechanism. Several utilities noted that the incentive attracted senior management attention, and therefore, DSM ascended the list of utility priorities. One utility noted that the incentive has helped some DSM skeptics recognize the benefits of DSM.

Other factors that contributed to the increased levels of DSM activity included a desire to improve customer service, particularly by utilities with high rates; the need to address environmental concerns,

---

[4] DSM activity has actually declined from mid-1980s levels at several New Jersey and Pennsylvania utilities. The primary reason for the decline is apparently the gradual phaseout of DSM programs that were begun following the accident at the Three Mile Island power plant. These programs were established to help defray rate increases necessitated by the high cost of replacement power. According to many of the New Jersey and Pennsylvania utilities, this trend will reverse in 1992.

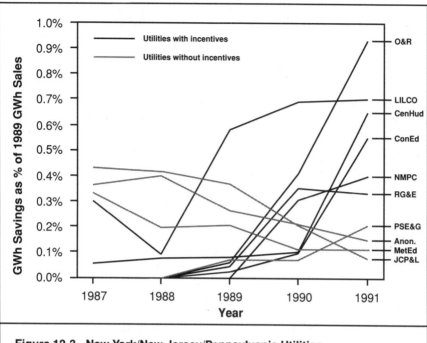

Figure 12-3. New York/New Jersey/Pennsylvania Utilities

including avoiding emissions and siting new facilities; and the fact that DSM was the least-cost resource. For example, one utility conducted a comprehensive study and was surprised to find that DSM was the most cost-effective resource. In addition, NYSEG credited its recently completed collaborative program-design process involving environmentalists and others.

In New York State, a number of different incentive mechanisms have been or are being used, including bonus payments linked with lost-revenue recovery, shared savings, and rate-of-return adders based on achieving specific goals. Since experience with each of these mechanisms is limited, no definitive conclusions can be made. However, several people noted that the shared-savings mechanisms tended to encourage commercial sector programs where the net benefits of DSM tend to be greatest. The mechanisms based on achieving specified kW and kWh goals were reported to result in increased emphasis on achieving these short-term goals—Long Island Lighting, with a kW goal in its initial incentive, emphasized kW savings, while Orange & Rockland, with a kWh goal in its second incentive, emphasized kWh savings. According to one intervenor organization, the lost-revenue recovery bonus in Long Island Lighting's first incentive, under which it could recover 120% of lost revenues, provided an incentive to increase, and perhaps even exaggerate, lost revenues.

In New York State there have been incentives for two years, crystallizing a number of issues. For example, three of the incentive mechanisms include penalties if DSM performance is below par. For several utilities this is troubling, but others think it will be relatively easy to achieve enough savings to avoid penalties. Commission staff and intervenors generally favored penalties, particularly for utilities with less than exemplary previous DSM efforts.

Most incentives have a cap on the maximum incentive that can be earned. A few utilities have reached their cap, noting that when the cap is reached they have little incentive to pursue additional DSM. Many utilities argued to either eliminate or raise the caps. Commission staff indicated that the optimum incentive level has not been resolved and will be revisited in the future.

Four of the incentives are paid only after an end-of-the-year reconciliation when the utility submits its estimates of energy and net resource savings to the Commission for approval. Several utilities indicated that they would like to receive at least a portion of the incentive earlier. A number of utilities, particularly those with shared-savings incentives, found the calculation process complicated.

Finally, many issues have arisen related to the specific shared-savings formulas used by the different utilities. These issues are discussed by Gallagher (1991) and will not be considered here.

West Coast. California's three largest investor-owned electric utilities receive incentives. Of the major private California utilities, only Pacific Power and Light does not receive an incentive.[5]

In Oregon, Portland General Electric receives an incentive, but Pacific Power and Light does not; however Pacific is allowed to ratebase its DSM investments. None of the Nevada utilities presently receives incentives.

The level of DSM activity at these utilities produces some interesting patterns. In the late 1980s, activity levels in California and Nevada were similar, but activity was more limited in Oregon (see Figure 12-4). In the early 1990s, activity increased enormously among the

---

[5] Pacific ratebases DSM investments. In addition, it is allowed to earn a return on investments in energy services, as described in Chapter 9. However, staff at the utility and intervenor organizations do not consider these to be much of an incentive.

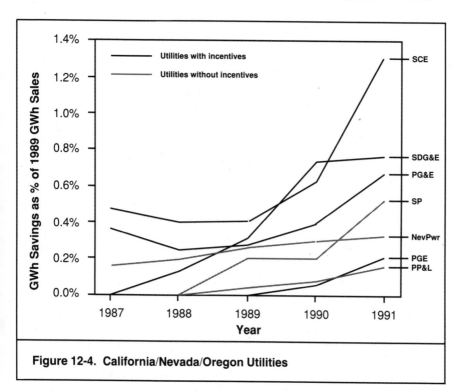

Figure 12-4. California/Nevada/Oregon Utilities

three large California utilities, with more modest increases among the other West Coast utilities.[6]

Utility, utility commission and intervenor staff interviewed in California attribute much of the increase in DSM activity to the incentives. All utilities noted that the incentives were important in coalescing support for DSM. Other factors also played a significant role, but the incentives were the catalyst that brought these different factors to management's attention.

The impact of these other factors cannot be minimized, including a statewide collaborative DSM planning process that developed the incentives as well as DSM program designs; regulator interest in DSM (regulators virtually ordered the utilities to participate in the collaborative process is an effort to jump-start utility DSM efforts); environ-

---

[6] An exception to this pattern is Sierra Pacific, where DSM activity increased substantially in 1990–91, primarily due to the Nevada Public Service Commission's decision to not approve the DSM portion of Sierra Pacific's 1989 resource plan. The revised plan, submitted in mid-1990, increased DSM staff threefold and spending fivefold.

mental concerns (public relations programs of major California utilities often emphasize an environmental theme); and customer service concerns (some customers wanted more DSM programs similar to those offered in the early to mid-1980s and then disbanded). In addition, some utilities realized that new capacity would be needed and that DSM was the most attractive option. Finally, in the case of San Diego Gas & Electric, concerns about high rates that had caused the utility to underspend its DSM budget were fading, and it again became acceptable to spend money on DSM.

In Oregon, the impact of the Portland General Electric incentive is apparently less dramatic, although it is too early to be sure. The amount of incentive available to the utility is fairly small (see Table 12-1), which could explain this finding. According to utility staff, a corporate commitment to delay new generating facilities had increased reliance on DSM as a resource. Still, the incentive reportedly has helped to accelerate and sustain this commitment. In addition, the incentive has helped to energize employees who work on DSM programs.

In California, there are two major issues: the level of incentive, and the role of evaluation in incentive determination.

Staff at the utility commission and several intervenors question if incentive levels are too high. These concerns particularly affect San Diego Gas and Electric which in 1990 was eligible for $16 million in incentives based on $17 million in DSM spending. Most of the San Diego incentive was based on a mechanism adopted in 1989 under which most of the DSM benefits accrue to the utility after meeting specified targets. This mechanism was developed when San Diego Gas & Electric was disinterested in DSM. The mechanism developed to deal with this included a penalty for poor DSM performance. Neither the utility nor the commission expected the utility to earn a bonus. For this reason, not much attention was given the bonus, with the result that it proved to be extremely generous. In fact, after intervenors and commission staff protested that the incentive was too high, in 1991 the Commission and the utility agreed to retroactively impose a cap on the incentive which cut the 1990 incentive by 42%.

Incentive levels are also a concern with Pacific Gas and Electric (PG&E). One observer felt that PG&E was trying to maximize its incentive in the short-term through cream-skimming, expecting that incentive levels would be scaled back in a few years.

Measurement of energy savings is also a significant issue in California where shared-savings incentives are based on engineering calculations made during program planning. During the first three years of program operations, each utility must measure the actual savings, and the results are used as the basis for engineering estimates made in future years. However, the measured savings are not used to adjust incentives earned for prior DSM programs. This lack of retroactive adjustments is popular with the California utilities since it assures that a given level of DSM activity will produce a specified incentive amount. Several observers consider the initial engineering estimates to be overly generous, but preparation of impact evaluations is proceeding slowly and thus no results were available to adjust the engineering estimates used for 1992 programs.

Conversely, the Oregon incentive is based on measured savings, and the Oregon commission is concerned about accurate evaluations. Utility staff worry about spending too much time and effort to develop overly precise numbers, diverting attention from more important DSM concerns.

In both California and Oregon, another important issue is whether shared-savings incentives should be based on utility costs or total resource costs (utility plus customer costs). The PG&E incentive is based only on utility costs, so the utility reportedly gives limited attention to customer costs. Conversely, the San Diego shared-savings incentive is based on total resource costs without regard to how much is paid by each party, with the result, according to observers, that the basic shared-savings incentive does not encourage the utility to control its costs.[7] Both the Portland General incentive and a new Southern California Edison shared-savings incentive approved in December 1991 give weight to both utility and total resource costs in their net-benefit formulas.

Finally, two other attributes of the San Diego and Portland incentives are worth noting. In San Diego, management instituted a system where a portion of the incentive the utility earns beyond a target is passed on to DSM employees. According to utility and commission staff, this arrangement inspired DSM employees, although utility and commission management wonder if employees are too interested in maximizing their personal gain. In Portland, in calculating the value of DSM benefits, a higher avoided cost value is used for the residential sector ($0.06/kWh) than for the commercial and industrial sectors ($0.05 and $0.04/kWh respectively) because residential loads are

---

[7] However, to address this problem, another aspect of the incentive mechanism rewards the utility for savings in program costs per unit of energy saved relative to budgeted amounts, and penalizes the utility for exceeding these targets. Observers disagree on the effectiveness of this cost control mechanism.

more closely aligned with the system peak. This arrangement provides a greater incentive to pursue residential sector programs than if the same avoided cost was used for all sectors.

Wisconsin. In late 1986, the Wisconsin Public Service Commission established an incentive mechanism for Wisconsin Electric Power Company (WEPCo) as part of an effort to replace a planned power plant with DSM. At the same time, the Commission began experimenting with different methods to promote DSM activity at other Wisconsin utilities, including a utility energy services program at Wisconsin Power & Light (WP&L) and a competition between Madison Gas & Electric (MG&E) and private DSM providers. Each of these efforts are described in Chapter 11.

Of these efforts, only the WEPCo program included a significant financial incentive, so it is the only Wisconsin program covered by the data analysis. Unfortunately, the data to analyze this four-year old incentive were not readily available. Based on limited data, DSM activity at WEPCo increased substantially after the incentive, while activity increased only modestly at WP&L at the same time.

In interviews with people involved in Wisconsin DSM efforts, it was generally agreed that the incentive was a significant contributor to the increased DSM efforts by WEPCo, although other factors were also involved, including long-range planning processes undertaken by both WEPCo and the state of Wisconsin, and a close working relationship between the Wisconsin Commission and the utilities it regulates (this relationship is described in Chapter 11). By 1991, according to Commission staff, the level of DSM activity is similar across all major Wisconsin utilities, although as a result of the incentive, WEPCo's efforts geared up quicker and sooner.

The WEPCo incentive was scheduled to last four years; hence it ended in December 1990. As of 1991, DSM efforts at WEPCo are going strong. WEPCo proposed a new incentive mechanism in 1990, but this was denied by the Commission, partly due to concerns about the specifics but also because the Commission believes that incentives should be intermittent, so that utilities and shareholders do not come to expect them. In lieu of an incentive mechanism for shareholders, in 1991 the Wisconsin Commission ordered WEPCo to establish a $500,000 pool to provide bonuses to WEPCo employees who help the utility achieve its DSM goals. This mechanism is described in Chapter 11.

Several lessons were learned from WEPCo's experience with incentives. First, the incentive was based on gross MW savings calculated using engineering estimates. The incentive did not promote kWh savings, or efforts to minimize program costs, limit free-riders, or maximize the amount of actual savings. WEPCO was somewhat concerned about these issues; however, its concern was not caused by the incentive. Second, accounting for savings proved to be time consuming for both the utility and the Commission, and caused a number of disputes. According to the utility, data are critical to determine the amount and cost of the DSM resource, and thus, even without the incentive, they continue to compile this information. Third, according to a Commission staff member, in the early years of the incentive, in an effort to achieve the MW savings threshold needed to trigger the incentive, WEPCo engaged in some "cream-skimming." However, after the "cream was skimmed" WEPCo concentrated on achieving more in-depth savings. WEPCo denies engaging in cream-skimming.

Colorado. In Colorado, Public Service of Colorado (PSCO) receives incentives for a few DSM programs, approved on a program-by-program basis. Currently some programs, particularly a third-party bidding program, are eligible for incentives, while others are not. Relative to CENTEL, a Colorado utility that does not receive incentives, DSM activity and targets at PSCO are substantially greater.

According to PSCO staff, the incentive has caught the eye of senior management, and DSM targets were increased. However, PSCO thinks the incentive is fairly small. Echoing this view, one outside observer noted that PSCO has yet to embrace DSM, and is still in the pilot program stage. PSCO responds that if it is to pursue additional DSM, it wants to be rewarded either with incentives or lost-revenue recovery. PSCO also notes that the Commission can order DSM programs, but it cannot order enthusiasm for them. Incentives, they claim, can help generate this enthusiasm.

Michigan. In 1990, a settlement agreement between Consumers Power and the Michigan Public Service Commission set forth nine DSM programs that Consumers would operate, and provided a small bounty-type incentive for four of the nine programs. These four programs accounted for approximately 75% of DSM expenditures. According to both Commission and utility staff, the incentive caught the utility's attention, and in combination with the Commission's desire for increased DSM activity, started a shift in attitude at the utility. In particular, according to Commission staff, design and implementation of targeted DSM programs improved.

However, this initial incentive was limited in amount, and the utility will not receive its full incentive until 1993. Also, the utility is not reimbursed for base revenues lost due to DSM programs. According to utility and Commission staff, the increase in DSM activity from the 1990 settlement and incentive, while significant, was modest.

In 1991 however, the Commission ordered Consumers to dramatically increase DSM activity, and instituted a new and much larger incentive/penalty mechanism. This new mechanism is an adjustment of the utility's overall rate of return—up to a 1% increase or a 2% decrease, depending on performance. As a result of this development, the utility sees DSM as a priority activity. However, this impact may be due to the potential penalty, not the potential reward since the potential penalty is twice the size of the potential reward. Also, the incentive rewards spending and cost-effectiveness up to a set expenditure level. When this spending requirement is met, there is no incentive for additional spending beyond what the Commission ordered.

Conclusions

Results from the data analysis and series of interviews indicate that providing utilities with financial incentives does, on average, have a positive and statistically significant impact on levels of utility DSM activity. Incentives attract senior management attention and help to quiet DSM skeptics within the utility.

However, many factors affect levels of DSM activity, including least-cost planning, environmental, customer-service, and regulatory-relations issues. Some utilities without incentives have substantially increased their levels of DSM activity, while other utilities with incentives have only marginally changed their DSM activity levels. Thus, the availability of incentives does not guarantee a strong DSM program and, in some cases, a strong DSM program can be developed without incentives. Incentives should be viewed as one element that can contribute to increased levels of DSM activity.

Experience with incentives is too limited to generate conclusions as to the relative merit of different approaches, the optimal amount of incentive, and whether incentives should be continuous or intermittent. The people interviewed for this study indicated that shared-savings incentives tend to promote commercial and industrial DSM programs more than residential programs, and said that data tracking for shared-savings programs is complex. According to the interviewees, bounty- and expenditure-based incentives are often simpler to administer than shared savings, but may send inappropriate signals. For example, bounty programs alone do not encourage efforts to improve program cost-effectiveness, and a kW-based bounty does not encourage kWh savings. Likewise expenditure-based incentives often do not encourage efforts to improve cost-effectiveness or savings. Even within the context of a shared-savings incentive, particular elements within the shared-savings formula can send signals that favor one type of DSM activity over another. For example, according to Gallagher (1991), incentive formulas based on savings achieved in the current year only tend to encourage short-term savings, while incentive formulas based on cumulative savings over a measure's lifetime tend to encourage long-term savings.

The number of utilities earning incentives is increasing dramatically each year. Our research indicates that incentives tend to work, and hence this trend should be encouraged. However, an evaluation of this type needs to be repeated in a few years when more data are available, including additional data about existing programs, and data on new incentives programs. Follow-up analyses should be based on statistical analyses of metered electricity consumption before and after each program, not the engineering estimates employed in this analysis, because incentive effectiveness should be based on actual savings. Such an analysis can verify or refute our preliminary conclusions and provide insight about the relative merits of different incentive approaches, the impact of different incentive levels, and how best to link incentives to actual savings.

Acknowledgments

We are indebted to the many people who provided data and agreed to be interviewed for this study. Much of the data was not readily available, with the result that staff at utilities and utility commissions often spent many hours gathering and interpreting the data requested.

Helpful comments on a draft of this work were provided by: Charles Budd, Consumers Power; Richard Cowart and Rick Weston, Vermont Public Service Board; James Gallagher, N. Y. Department of Public Service; Howard Geller, ACEEE; James Hagerman, Oregon Public Utility Commission; Eric Hirst, Oak Ridge National Laboratory; Mary Kilmarx, R.I. Public Utilities Commission; Philip Koebel, Southern California Edison; Marty Kushler, Michigan Public Service Commission; John Locher, Detroit Edison; Corey Pettett, Wisconsin Electric; Jonathan Raab, consultant; David Robison, Pacific Power & Light; Laura Rooke, Portland General Electric; Carol White, Eastern Utilities Associates; David Wooley, Pace University Center for Environmental Legal Studies.

Funding for this work was provided by the New York State Energy Research and Development Authority.

References

Blackmon, Glenn. 1991. "Conservation Incentives: Evaluating the Washington State Experience." *Public Utilities Fortnightly*. January 15. Pp. 24–27.

Gallagher, James. 1991. "DSM Incentives in New York State: A Critique of Initial Utility Methods." In *Proceedings 5th National Demand-Side Management Conference*. CU-7394. Palo Alto, Calif.: Electric Power Research Institute. Pp. 220–226.

Hoel, Paul G. 1971. *Introduction to Mathematical Statistics*. Fourth edition. New York, N. Y.: John Wiley & Sons, Inc.

Massachusetts Electric. 1991. *1990 DSM Performance Measurement Report*. Westborough, Mass.: New England Electric System.

Raab, Jonathan D., and Martin Schweitzer. 1992. *Public Involvement in Integrated Resource Planning: A Study of Demand-Side Management Collaboratives*. ORNL/CON-344. Oak Ridge, Tenn.: Oak Ridge National Laboratory.

List of People Who Were Interviewed and/or Provided Data

Ken Anderson and David Robison, Pacific Power & Light, Portland, Oreg.

John Angeli and Phil Turner, United Illuminating, New Haven, Conn.

Bill Atzl and Jim Cuccaro, Orange & Rockland, Pearl River, N.Y.

Janet Besser, New Hampshire Public Utilities Commission, Concord.

Calvin Birge, Pennsylvania Public Utility Commission, Harrisburg.

Eric Blank, Land & Water Fund of the Rockies, Boulder, Colo.

Bill Branch and Greg Lambert, Sierra Pacific, Reno, Nev.

Charles Budd, Consumers Power, Jackson, Mich.

Arthur Canning and Philip Koebel, Southern California Edison, Rosemead, Calif.

Phil Carver, Oregon Department of Energy, Salem.

Ralph Cavanagh, Natural Resources Defense Council, San Francisco.

Joe Chaisson, Conservation Law Foundation, Boston.

Connie Colter and Jim Hagerman, Oregon Public Utility Commission, Salem.

Richard Cowart and Rick Weston, Vermont Public Service Board, Montpelier.

Scott Cunning and Bill VinHage, New York State Electric & Gas, Binghamton.

Brian Daly, Fred Link, and Joe Polasky, Public Service Electric & Gas, Newark, N.J.

Paul Decotis and Peter Smith, New York State Energy Office, Albany.

John Dillon and Stephen Purtusiello, Consolidated Edison, New York, N.Y.

George Dunn and Mark Quinlin, Connecticut DPUC, New Britain.

Jennifer Fagan, Wisconsin Power & Light, Madison.

Tom Fogg and Marty Morse, Rochester Gas & Electric, Rochester, N.Y.

James Gallagher, Bill Mills, and Marsha Walton, New York Department of Public Service, Albany.

John Glusko, Central Hudson Gas & Electric, Poughkeepsie, N.Y.

Suzanne Halpin, Ray Plaskon, and Steve Maslak, Long Island Lighting Co., Hicksville, N.Y.

John Hartnett, Niagara Mohawk, Syracuse, N.Y.

Liz Hicks, Meredith Miller, Tim Stout, Joe Wharton, and Dean White, New England Power Service, Westborough, Mass.

Tom Holderfield and John Moore, Public Service of Colorado, Denver.

Jim Kaul and Carol Stemrich, Wisconsin Public Service Commission, Madison.

Henani Kekuna and John Lucas, Pennsylvania Power & Light, Allentown.

Kathy Kelly, Boston Edison, Boston.

Mary Kilmarx, Rhode Island Public Utilities Commission, Providence.

Lee Kline, Jersey Central Power & Light, Morristown, N.J.

Marty Kushler, Michigan Public Service Commission, Lansing.

Brad Latham, Rich Soderman, and Kathy Thayer, Northeast Utilities, Hartford, Conn.

John Locher, Detroit Edison, Detroit, Mich.

Scott MacNevin and Mort Zajac, Commonwealth Electric, Wareham, Mass.

Theo MacGregor, Massachusetts Department of Public Utilities, Boston.

Michael Messenger and Penny Purcell, California Energy Commission, Sacramento.

Louis Milford, Conservation Law Foundation, Montpelier, Vt.

Bill Miller and Dan Quigley, Pacific Gas & Electric, San Francisco.

Mona Mosser, New Jersey Board of Regulatory Commissioners, Newark.

Jerrold Oppenheim, Massachusetts Office of the Attorney General, Boston.

Corey Pettett, Wisconsin Electric Power Co., Milwaukee.

Jonathan Raab, Energy Consultant, Jamaica Plains, Mass.

Bill Riggs, CENTEL Electric, Pueblo, Colo.

Laura Rooke and Keith White, Portland General Electric, Portland, Oreg.

Deborah Ross, Washington Utilities and Transportation Commission, Olympia.

Gary Schmitz and Morey Wolfson, Colorado Public Utility Commission, Denver.

Don Schultz, California Public Utilities Commission, Sacramento.

Stephen Spaar, Metropolitan Edison, Reading, Pa.

Mike Weedall, Green Mountain Power, So. Burlington, Vt.

Carol White, Eastern Utilities, West Bridgewater, Mass.

Yole Whiting, San Diego Gas & Electric, San Diego.

Chuck Willems & Ron Zanoni, Nevada Power, Las Vegas.

David Wooley, PACE University Center for Environmental Legal Studies, White Plains, N.Y.

Chapter 13

DSM Incentive Mechanisms: Comparative Assessment and Future Directions

David R. Wolcott and Steven M. Nadel

Introduction

In this concluding chapter, we examine the strengths and weaknesses of each of the demand-side management (DSM) incentive mechanisms presented in the book. We compare and summarize the incentive mechanisms in terms of their abilities to address the different design criteria introduced in Chapter 1. We conclude with a look into the future regarding the issues that will most likely shape the continuing evolution of regulatory incentives for DSM.

Comparison of DSM Incentive Mechanisms

We must sound a note of caution regarding any attempt to rank order the DSM incentive mechanisms. Such a simplistic approach would not be very useful in understanding the subtle differences between the mechanisms and how they interrelate. The development and use of DSM incentive mechanisms depend on the specific goals of regulators and motivations of utilities involved in the process. Therefore, we will not try to define the "best" mechanism, which in most cases will probably be a combination of mechanisms.

For the sake of clarity in this comparative analysis, we use some shorthand to identify the DSM incentive mechanisms treated in this book as follows:

- The Revenue Decoupling Mechanisms (**RDM**) such as:
  - Electric Rate Adjustment Mechanism (**ERAM**) (Chapter 3)

- Revenue per Customer (**RPC**) (Chapter 4)
- **Ratebasing** (Chapter 5)
- Adders such as the **Bounty** and **Markup** (mentioned in Chapter 2)
- **Shared Savings** (Chapter 6)
- Return on Equity (**ROE**) **Adjustment** (Chapter 7)
- **Bill Indexing** (Chapter 8)
- **Energy Services** (Chapter 9)

The criteria for designing DSM incentive mechanisms were introduced in Chapter 1 and then touched upon in each of the chapters on the different mechanisms. The following assessment of the strengths and weaknesses of DSM incentive mechanisms is presented in terms of these criteria. We caution the reader not to view each criterion as equally important; nor will any one incentive mechanism address all criteria. A combination of mechanisms may be required to meet particular regulatory and utility goals.

Recovery of DSM Program Costs

All participants in the integrated resource planning (IRP) process agree that recovery of DSM program costs is required in order to obtain the enthusiastic support of utility management to aggressively pursue IRP goals. This view holds that utilities should be able to recover all allowable direct costs to design, implement, and evaluate DSM programs. These costs include the incentives paid to customers (such as rebates) to motivate their participation in the programs, payments to vendors (such bids by energy service companies) to compensate them for implementing programs, and other program-related expenses such as advertising, labor and administration.

States that have adopted incentive mechanisms typically allow for the recovery of DSM program costs. In most cases the cost recovery is not inherent in the incentive mechanism but is provided through another complementary process for this purpose, such as a surcharge or a deferral mechanism. The exceptions in which DSM program cost recovery is inherent to the mechanism include **Energy Services**, **Ratebasing**, and **Markup**.

Recovery of Lost Revenues

A major disincentive to DSM is the lost revenue (associated with unavoidable fixed costs) from foregone sales of electricity that customers don't buy because of conservation. Absent decoupling of revenues from sales (see the following point), utilities should be able to recover the lost revenue on fixed costs. Variable costs (primarily for fuel) are not incurred and are therefore not "lost."

The **RDMs** inherently provide for the removal of the disincentive represented by lost revenues from DSM. In all other cases in which lost revenues are accounted for and recovered, it is through the explicit recognition of them as allowable costs.

Decoupling Revenues from Sales

Decoupling the direct relationship between a utility's revenues and its sales volume removes the utility's short-run incentive to sell additional electricity. While decoupling should eliminate the issue of the recovery of lost revenues from DSM, it also has a much broader impact on utility operations.

The **RDMs** by definition explicitly provide for the decoupling of revenues from sales. **Bill Indexing** can be described as an indirect decoupling mechanism. While there may be every intention to decouple, there is no guarantee of that outcome. The other mechanisms do not decouple, although they can be combined with a decoupling mechanism as part of a comprehensive regulatory reform package.

Some DSM advocates and regulators believe that decoupling is a fundamental fix to the deficiencies of traditional regulation. However, other DSM advocates do not necessarily see decoupling itself as a panacea because it does not provide a positive incentive to promote DSM (see the following point). Furthermore, since decoupling shifts risk from utilities to ratepayers, it can have the side effect of making utilities indifferent to issues that could raise rates (e.g., cogeneration bypass) or degrade customer service. Utilities are less quick to embrace decoupling because it removes a potential avenue to profitability.

Provision of Positive Incentives

A positive incentive may be required to motivate utility management to overcome the perceived risk of implementing DSM programs. Regulators increasingly agree that utilities should earn increased profits, at least initially, for successfully accomplishing their goals in IRP (e.g., implementing a resource plan with the lowest societal costs) and DSM (e.g., acquiring all cost-effective DSM resources).

Five of the mechanisms presented in this book provide positive incentives. Each mechanism provides a bonus in a different way. The **RDMs** do not provide a positive incentive; however, most recent implementations of **ERAM** and **RPC** include a complementary performance-based bonus.

Among the five positive incentive mechanisms, there are no inherent differences in the amount of bonus provided. Instead, the value of the incentive depends on several other factors that can be built into any mechanism.

The primary factor is the magnitude of the incentive that regulators allow. Regardless of the mechanism, regulators generally decide what an appropriate incentive amount is for an expected level of utility performance. Once this amount is determined, it can be earned on a **Shared Savings, Bounty, Markup, Ratebasing,** or **ROE Adjustment** basis.

A second factor is whether thresholds are established that must be exceeded before incentives are earned. Some states allow incentives to be earned on the first kWh saved while other states require that a threshold amount of savings must be achieved before an incentive begins to accrue (other states even provide penalties if savings are particularly low).

A third factor is whether caps are established to limit the maximum incentive that can be earned. Caps can keep incentives from becoming excessive, but once the cap is reached a utility has little incentive to continue pursuing the desired behavior.

A final factor is the amount of time over which a bonus is earned. A bonus today is better than a bonus tomorrow, considering both the time value of money and the uncertainties of regulatory continuity.

Performance-Based and Measurable

Incentive mechanisms can reward and/or penalize the utility's performance in accomplishing IRP and DSM goals. A direct linkage between an incentive mechanism and a utility's performance in successfully acquiring cost-effective DSM resources provides the clearest possible signal to DSM managers. Incentives should increase as actual energy savings increase and vice versa. Such linkage is generally considered to be necessary to motivate specific utility behavior that is closely associated with IRP goals.

The **Shared Savings** and **Bounty** mechanisms inherently reward utility performance in accomplishing DSM. So does **ROE Adjustment**, provided the adjustment is tied to DSM performance. **Shared Savings** and Orange & Rockland's **ROE Adjustment** reward cost-effective DSM performance, while the **Bounty** simply rewards energy savings without attention to program cost, which could compromise cost-effectiveness. Depending on how it is structured, **Energy Ser-**

vices can also reward performance by allowing the utility to keep its share of the savings achieved. The other incentive mechanisms presented in this book are not performance-based.

One benefit of performance-based incentive mechanisms is that DSM program performance lends itself to measurement techniques. However, there are tradeoffs to the extent that the determination of energy savings may require expensive monitoring and impact evaluation. As discussed in Chapter 10, impact evaluation brings both credibility and knowledge about making programs more effective, but at a cost. Performance must be objectively determined without excessive expenditures of time and money. Otherwise, marginal DSM programs may not pass cost-effectiveness tests.

The criteria of "performance-based and measurable" are not inherently linked, as illustrated by the **Markup**, which is very measurable but not performance-based. Also, there can be differences in what is being measured. For example, the **Bounty** measures savings, **Energy Services** measures savings and utility costs, and **Shared Savings** measures savings, utility costs, and customer costs (in some cases).

Understandable, Administrable, Predictable

DSM incentive mechanisms must be readily understood by all stakeholders (e.g., utilities, regulators, intervenors, ratepayers) in the regulatory process and should not be overly complex or difficult to administer. It is also useful to know in advance that certain utility behavior will produce a particular effect.

An incentive mechanism is understandable if utility implementors and other participants in the regulatory process can easily grasp its mechanics and implications. In this context, the best mechanism is one that is readily understood by ratepayers. Three mechanisms are all fairly straightforward in this regard: **Bounty** depends on energy savings, **Markup** depends on program expenditures, and **Shared Savings** depends on the difference between savings and expenditures. Each of these concepts are easy to understand. On the other end of the spectrum, the concepts of attrition and balance accounts under **ERAM** can be difficult for ratepayers to understand.

An incentive mechanism is most easily administrable if the utility and regulators face a tractable process of data collection and presentation within the context of normal regulatory proceedings. Mechanisms such as **Ratebasing, Bounty** and **Markup** are inherently administrable in that they fit easily into the existing regulatory process. While **Shared Savings** is easy to understand, it can be more difficult to administer because of the complex measurements and calculations that are required, particularly when customer costs are included in the shared-savings formula. It remains to be seen if **Bill Indexing** and **Energy Services** are easy to administer or not, although they probably have more in common with **Shared Savings** than with other mechanisms.

Predictability refers to the extent to which certain utility actions result in expected incentives. This is *not* the same issue as whether DSM savings are predictable given certain utility programs. **Bounty** and **Markup** are the most predictable because they provide a fixed incentive for savings or program expenditures, respectively, without recourse to another variable. Alternatively, the **Shared Savings** incentive is earned as a percentage of total resource savings that are based on both DSM savings and program expenditures. **Ratebasing** is relatively predictable. Although there is some risk to the utility of under-recovery of costs in rate cases, that risk is probably low. **Bill Indexing** is perhaps the least predictable mechanism because the incentive earned depends on the behavior of the target utility relative to other utilities.

Evaluation methods can have an interesting effect on how predictable an incentive can be. As discussed in Chapter 9, the use of engineering estimates to determine *ex ante* levels of program savings gives a utility a great deal of control over the incentive payment it will receive. Alternatively, *ex post* levels of program savings determined by impact evaluations are much less certain and will result in less predictable incentives. Nonetheless, the predictability benefits of engineering estimates must be balanced against the greater certainty of DSM savings provided by impact evaluation. Such certainty applied to load and resource forecasts is important in the IRP process.

Ability to Influence Other Outcomes

DSM incentive mechanisms that motivate utility pursuit of IRP goals can have collateral consequences, both positive and negative. A desirable outcome might be that a mechanism minimizes the costs of DSM to society and program nonparticipants. Undesirable outcomes might include situations in which utilities can indulge in gaming techniques or solely pursue cream-skimming DSM measures. Naturally, the ideal DSM incentive mechanism provides clear signals to motivate utilities to pursue DSM while maximizing other desirable outcomes and minimizing unintended negative consequences.

In terms of minimizing DSM costs to society and nonparticipants, the **Shared Savings** mechanism minimizes utility or societal costs depending on the formula that is used. Orange & Rockland's **ROE Adjustment** tends to minimize program costs. For **ERAM** and **Bill Indexing**, this benefit can be blunted somewhat if the utility cuts back on customer services to achieve cost savings. **Bill Indexing** minimizes average customer bills while **Energy Services** minimizes costs to nonparticipants. The other mechanisms don't provide any inherent process of minimizing costs to ratepayers. **Ratebasing** and the **Markup** may encourage DSM program cost maximization or "goldplating."

Gaming refers to a situation in which participants exploit opportunities in the regulatory process to achieve unintended consequences. The classic example of gaming is where a utility underestimates load growth in a rate case so that it can work hard to exceed sales estimates before the next rate case to achieve additional profits. The **RDMs** tend to avoid gaming since decoupling removes the opportunity to exploit the difference between forecasted and actual costs between rate cases. **RPC** could be an exception to the extent that a utility could game the customer count. Alternatively, lost revenue recovery mechanisms are generally susceptible to gaming as discussed in Chapter 1. All the other mechanisms rely on traditional regulation (i.e., rate cases) for their implementation and therefore neither especially invite nor discourage gaming.

Cream-skimming is the practice of implementing only those DSM measures that have the shortest payback times—e.g., efficient lighting. The **Markup** discourages cream-skimming because promotion of high-cost, long payback measures will generally increase the amount of incentive earned. With the **Bounty** and **Shared Savings** incentives, cream-skimming can be a problem in the short-term because the incentive earned per dollar invested is maximized by promoting measures with rapid paybacks. However, because incentives can be earned on any cost-effective measure, there is a positive incentive to pursue longer payback measures, although many of these measures may only be pursued in the long-term after the more lucrative measures have been tapped.

Strengths and Weaknesses of DSM Incentive Mechanisms

We draw from the preceding discussion to present the strengths and weaknesses of each DSM incentive mechanism in Table 13-1. We recognize the risk of creating a "scorecard" that could be used to rate each incentive mechanism by design criterion to come up with winners

Table 13-1. Summary of Different Incentive Mechanisms								
	ERAM	RPC	Rate-basing	Bounty	Markup	Shared Savings	ROE Adj.	Bill Indexing	Energy Services
Cost recovery	N	N	Y	N	Y	N	N	N	Y
Lost revenue recovery	Y	Y	N	N	N	N	N	N	N
Decoupling	Y	Y	N	N	N	N	N	N	N
Positive incentives	N	N	M	Y	Y	Y	Y	Y	Y
Performance-based	N	N	N	Y	N	Y	M	M	M
Measurable	NA	NA	Y	Y	Y	P	P	P	P
Understandable	M	M	Y	Y	Y	Y	Y	M	Y
Administrable	P	P	Y	Y	Y	P	P	M	M
Predictable	Y	Y	Y	Y	Y	P	P	N	P
Minimizes societal cost	N	N	N	N	N	M	M	M	N
Minimizes non-participant cost	NA	NA	N	N	N	N	N	M	Y
Discourages gaming	Y	Y	M	M	M	M	M	M	M
Discourages cream-skimming	N	N	N	M	Y	M	M	N	M

Key:
Y = yes; P = probably; M = maybe; N = no; NA = not applicable.

and losers. Once again, we caution the reader to avoid such a side-by-side comparison of mechanisms without reflecting upon the regulatory goals and utility motivations of a particular situation. An additional caveat is that the following summary of each incentive mechanism is abbreviated and certainly not exhaustive. The reader should consult the chapters on each mechanism for a fuller treatment of these issues.

Electric Rate Adjustment Mechanism (ERAM): inherently removes the disincentive of lost revenues; explicitly provides direct decoupling of revenues from sales; can include a complementary performance-based bonus; can be difficult for ratepayers to understand;

can make utilities indifferent to issues that could raise rates or degrade customer service; and reduces gaming since decoupling removes the opportunity to exploit the difference between forecasted and actual costs between rate cases.

Revenue Per Customer (RPC): inherently removes the disincentive of lost revenues; explicitly provides direct decoupling of revenues from sales; can include a complementary performance-based bonus; is probably easier to administer than **ERAM**; and avoids gaming since decoupling removes the opportunity to exploit the difference between forecasted and actual costs between rate cases, although could invite gaming the customer count.

Ratebasing: inherently allows for DSM program cost recovery; can provide a positive incentive since the adder allows a utility to earn a bonus return as a percent of DSM program expenditures; is inherently administrable in that it fits easily into the existing regulatory process; is relatively predictable, but may encourage goldplating.

Bounty: provides a positive incentive that is performance-based as a function of the amount of DSM savings; is understandable and easy to administer; is predictable since the incentive is based on savings; does not encourage cost-effectiveness; and can present a cream-skimming problem in the short-term.

Markup: inherently allows for DSM program cost recovery; provides a positive incentive as a percentage adder on program costs; is measurable although not performance-based; is understandable, easy to administer and predictable; may encourage goldplating; and discourages cream-skimming because promotion of high-cost, long payback measures will generally increase the amount of incentive earned.

Shared Savings: provides a positive incentive based on the amount of DSM delivered by the utility; is performance-based as it inherently rewards utility performance in accomplishing DSM; is understandable; may be difficult to administer because of the complex measurements and calculations that are required (particularly when customer costs are included in the shared savings formula); is somewhat predictable especially when engineering estimates are used; minimizes utility or societal costs (depending on the formula which is used); and can present a cream-skimming problem in the short-term.

ROE Adjustment: inherently provides a positive incentive by applying a percentage bonus from the entire rate base; can be performance-based by tying the incentive to the amount of DSM saved; and tends to minimize program costs if the incentive is tied to net resource savings.

Bill Indexing: provides indirect decoupling of revenues from sales; may be difficult to administer; is unpredictable; and minimizes average customer bills, although utility may cut back on customer services to achieve cost savings.

Energy Services: inherently allows for DSM program cost recovery; can be performance-based if the utility is allowed to keep its share of the savings achieved; is measurable; may be difficult to administer; and minimizes costs to nonparticipants.

Future Issues in DSM Incentive Design

We conclude with a discussion of emerging issues in DSM incentive regulation as they may play out in the future. Considering all of the recent developments in regulatory incentives, the following issues appear likely to figure prominently in the implementation of DSM programs during the next few years.

Importance of Evaluation

As DSM assumes a larger role in many utilities' resource portfolios, greater emphasis is being given to careful planning and execution of evaluation studies. Impact evaluations, which examine DSM programs' true benefits and costs, become doubly important when the results serve as input to the determination of a utility's DSM bonus (or penalty).

In several instances, provision of DSM incentives has been contingent on a utility commitment to a detailed plan to evaluate its programs. The California collaborative established a precedent in this area with publication of evaluation protocols in proposing DSM incentives (California Collaborative 1990). The quality of utilities' evaluation efforts is likely to receive much greater scrutiny in regulatory proceedings, both when new incentive mechanisms are proposed and when bonuses are awarded based on claimed performance.

Treatment of Externalities

The environmental effects of energy production and consumption are increasingly being considered in utility resource planning decisions. By 1990, seventeen state public utility commissions were considering environmental externalities (benefits and costs not directly reflected in prices or rates) in utility regulation, although very few are doing so quantitatively (Cohen et al. 1990). Socioeconomic externalities are also being considered in some jurisdictions. For example, Nevada has adopted IRP rules that give consideration to employment and other regional economic impacts of resource decisions (Wiel 1991). In general, treatment of environmental externalities in planning favors DSM resources because they substitute for emissions-producing combustion processes for generation and thus are considered environmentally benign.

DSM incentives can be a vehicle for incorporating externalities since a utility's bonus or penalty can be related, in part, to the environmental consequences of its actions. This approach has been adopted in New York where an environmental externality value of $.014 per kWh is added to the benefits of DSM in the calculation of incentives (Gallagher 1991). Therefore, a New York utility earning a 10% **Shared Savings** bonus would receive an additional $.0014 per kWh. Incentive regulations in New Jersey operate similarly (New Jersey Board of Regulatory Commissioners 1991).

Federal Interest in DSM Incentives

Although regulation of retail utilities is largely the province of the states, Congress has expressed interest in DSM incentives and encouraged their use in legislation. The Clean Air Act Amendments of 1990 authorized the award of certain bonus emissions allowances (beginning in 1992) to utilities that invest in DSM and renewable resources before emissions caps take effect. For a utility to be eligible for such bonuses its public utility commission must have adopted reforms to ensure that the utility's net income is not reduced by DSM—i.e., a decoupling and/or DSM incentive mechanism must be in effect (Markey and Moorhead 1991).

Several energy and utility policy bills introduced in Congress in 1991 would go further, amending the Public Utility Regulatory Policies Act (PURPA) to add DSM incentives to a number of regulatory policies that states should consider. If enacted, such legislation would probably accelerate the adoption of DSM incentives.

Incentives for Publicly Owned Utilities

Regulatory incentives have thus far been limited to investor-owned utilities because the impact of DSM on company profitability is of major concern to these companies. However, 23% percent of electricity sales to end-users in the U.S. are by publicly owned utilities, and these utilities are also often concerned about the impact of DSM programs on revenues and rates. For example, some municipalities seek to make a "profit" on electricity sales to help fund municipal services, thereby keeping taxes down. Furthermore, publicly owned utilities (being part of a political system) are concerned about consumer reaction to rate increases, even if they are to pay for DSM programs that on average decrease utility bills. In addition, just as private-sector utility managers may be motivated in part by the desire to increase company size (as measured by sales or revenues), so too are some public-sector utility managers. Mechanisms to recover DSM program costs and avoid lost base revenues (through forward-looking sales and revenue planning and/or some type of annual true-up process) could address these concerns. In addition, other mechanisms, such as incentives for utility managers (similar to the Wisconsin Electric incentive described in Chapter 11) may be appropriate. This issue merits further research and perhaps some creative solutions.

Incentives for Other Least-Cost Resources

If regulation should be reformed to truly make a utility's least-cost plan its most profitable resource plan, then it is appropriate to examine the economic incentives and disincentives affecting the selection of *all* resources (both demand and supply) that could be part of the least-cost mix. Conceptually, there is a small gap between incentives for DSM and incentives for other resources that entail unfamiliar risks or do not ordinarily contribute to utility profitability.

Virtually all of the recent incentive activity, however, has been focused exclusively on DSM. One exception was a 1990 proposal by Central Maine Power (CMP) to receive a **Shared Savings** incentive not just on DSM but also on purchased power contracts, the costs of which are ordinarily passed through to ratepayers with no return to shareholders. CMP argued that the current system does not encourage aggressive actions to seek out the least costly purchase options (Central Maine Power 1990). Its proposal was eventually withdrawn in favor of a collaboratively developed proposal for DSM incentives, so the issue was never formally considered by the Maine public utility commission. However, CMP's proposal could anticipate the day that incentives for the acquisition of independent power, perhaps including renewable energy resources, will appear on the regulatory agenda. Incentives may also be applied to other operating efficiencies that utilities can pursue beyond the selection of new resources, such as repowering decisions and improvements in the operation of the transmission and distribution system.

How Much Incentive Is Needed?

After a utility commission decides to offer incentives, one of the next major decisions it faces is how large to make the incentive. On the one hand, the incentive should be large enough to motivate utility management to pursue the desired objectives. On the other hand, the incentive should not be so large to provide excess profits to utilities at ratepayer expense. While most parties to these discussions agree on these general guidelines, a consensus has yet to emerge on how much is "enough" without being "excessive."

At one extreme in this debate is the view presented in Chapter 11 that no incentive is needed to promote DSM because the current system provides sufficient incentive. This view is grounded in historical data suggesting that on balance, shareholders of fast-growing utilities have fared worse than shareholders of slow-growing utilities. The authors claim that investors recognize this phenomenon and therefore do not expect incentives for DSM that can curb fast growth and the attendant financial problems.

However, the view that no incentives are needed is predicated on a regulatory environment similar to the one in Wisconsin which includes such features as a forward-looking test year, annual rate cases, proactive commission involvement in strategic decisions, and the use of employee incentives for DSM and competition-like mechanisms that reward good performance. Where these features are not adopted, even the authors of Chapter 11 would probably agree that decoupling and incentive mechanisms may be necessary.

At the other extreme of this debate is the argument being discussed within some utilities that incentive amounts should be pegged to the "opportunity cost" of choosing DSM in lieu of generation investments. By choosing DSM, the utility may be foregoing an opportunity to add to its rate base and thus increase earnings. Even if the DSM is rate-based, the size of the DSM investment would be less than the foregone supply-side investment, so DSM's potential contribution to earnings would be less. For example, if we roughly estimate that the average DSM measure costs half as much as its supply-side equivalent, according to the opportunity cost argument, the return on the DSM investment would have to be double the return on the supply-side investment for the utility to be financially indifferent to the two choices.

Most parties to this debate espouse a view between these two extremes, which still allows for a large range of positions. An illustration of the range that is now being debated is provided in Table 12-1 (Chapter 12) which shows that annual incentives now being earned by utilities range from a low of 0.03% to a high of 0.94% of gross revenues. In coming years, utility commissions are likely to devote increasing attention to the question of "how much is enough?" Extensive debate and research are likely on this issue before a consensus begins to emerge.

Rate Impacts

In some states, large industrial customers have expressed concerns about how DSM programs affect electric rates (ELCON 1990). Rates can increase because DSM program costs must be recouped and because DSM programs reduce sales, leaving fewer revenues from which fixed costs can be recovered. On the other hand, DSM programs generally reduce customer bills because consumption decreases by more than rates increase. However, for customers who do not participate in DSM programs, consumption does not change and rate increases mean bill increases. Rate increases due to DSM are modest, in the range of 0.5% to 3.0% (Hirst 1991). But if a large industrial customer does not participate in DSM programs, such rate increases can raise annual bills by hundreds of thousands of dollars.

DSM incentives enter into the picture in two ways. First, incentives raise DSM costs and hence increase rate impacts. These effects are likely to be modest. For example, if a utility's DSM programs raise rates by 2% and an incentive equal to 10% of program costs is earned, rates will increase by only an additional 0.2%. Secondly and more importantly, incentives can encourage utilities to increase DSM spending, which in turn adds to rate impacts. Thus, if a utility doubles DSM expenditures, rates will tend to increase proportionately. Alternatively, increased spending will often result in increased customer participation in DSM programs, leaving fewer nonparticipants who will see their bills increase.To the extent that large industrial customers have opposed DSM programs on rate impact grounds, the debate has spilled over into proceedings on regulatory incentives. In order to contain the problem, it may be necessary for utilities to provide expanded DSM program offerings targeted specifically at the needs of industrial customers and to experiment with DSM programs in which participating customers pay the full cost of the programs (as described in Chapter 9).

The Effectiveness of DSM Incentives

The final and perhaps most important issue is a simple one: do DSM incentives work and which approaches work best in which situations? Given the multitude of factors affecting utility behavior and performance, isolating the effects of incentives is extremely difficult. Several years' worth of data may be needed before clear conclusions can be drawn. In the meantime, debates on the merits of incentives will undoubtedly continue. Questions that will be asked include:

- How large must incentives be to motivate aggressive DSM performance? How much is enough?

- Should incentives reward average levels of DSM performance or only exceptional performance?

- If incentives work now, will they continue to be effective, or will they lose their power as the novelty wears off?

- Should bonuses be phased out as utilities become more familiar with DSM? After incentives have "jump-started" utility DSM programs, should they be removed because utilities should be carrying out these programs as part of their obligation to serve, without the need for "bribes?" Should some aspects of incentive mechanisms, such as decoupling or lost revenue recovery, become permanent features of regulation?

The development of regulatory incentives for demand-side management is in its infancy as a concept that appears to have caught the attention of the public utility commissions of the United States. Even though we have had almost a decade of experience with ERAM in California, it is only in the last two to three years that we have witnessed the shared-savings and bonus mechanisms that have captured the hearts and minds of utility executives and regulators nationwide. No doubt the information provided in this book will need revising in a few years by virtue of the many surprises ahead as we all learn from the practical implementation of what heretofore has been abstract theory.

There is, however, one fact that is impossible to deny: the world of utility regulation has fundamentally changed, and there is no going back. For years government exhorted and cajoled utilities to undertake measures to increase energy efficiency in their customers' facilities with only limited success. Now utilities are designing and implementing programs that are enthusiastic, aggressive and imaginative efforts to accomplish that very goal. Clearly, the regulatory incentives that allow utilities to recover their costs and earn a profit through DSM have had an impact.

References

California Collaborative Process. 1990. *An Energy Efficiency Blueprint for California. Appendix A: Measurement Protocols for DSM Programs Eligible for Shareholder Incentives.* January.

Central Maine Power Company. 1990. Revised proposal submitted in Docket No. 90-085. August 1.

Cohen, S.D., J.H. Eto, C.A. Goldman, J. Beldock, and G. Crandall. 1990. *A Survey of State PUC Activities to Incorporate Environmental Externalities in Electric Utility Planning and Regulation.* Washington. D.C.: National Association of Regulatory Utility Commissioners. May.

ELCON. 1990. *Profiles in Electricity Issues: Demand-Side Management.* Washington, D.C.: Electricity Consumers Resource Council. December.

Gallagher, James. 1991. "DSM Incentives in New York State: A Critique of Initial Utility Methods." *Proceedings of 5th National Demand-Side Management Conference*, Boston, July 30–August 1.

Hirst, E. 1991. *The Effects of Utility DSM Programs on Electricity Costs and Prices.* ORNL/CON-340. Oak Ridge, Tennessee: Oak Ridge National Laboratory. November.

Markey, Edward J. and Carlos J. Moorhead. 1991. "The Clean Air Act and Bonus Allowances." *Public Utilities Fortnightly* 127. May 15.

New Jersey Board of Regulatory Commissioners. 1991. Rules adopted in Docket No. EX90040304. November 4.

Wiel, Stephen. 1991. "Nevada Adopts Clean Power Rule." Presented to Conference on Demand-Side Management and the Global Environment, Arlington, Va, April 22.

About the Editors

Steven M. Nadel is a senior associate with the American Council for an Energy-Efficient Economy where he directs ACEEE's work on utility and equipment issues. This work focuses on utility demand-side management efforts and on the efficiency of appliances, lighting, and motor systems. Prior to joining ACEEE, he spent two years planning and evaluating electricity conservation programs for the New England Electric System. He is the author or coauthor of more than 50 publications on energy issues.

Michael W. Reid is a project director with Barakat & Chamberlin in Washington, D.C.. He works with the firm's utility clients on demand-side management programs and regulatory strategies. He has directed projects on DSM cost recovery and incentives for numerous utilities, the Edison Electric Institute, the Electric Power Research Institute, and the National Association of Regulatory Utility Commissioners. Prior to joining Barakat & Chamberlin in 1989, he was deputy director of the Alliance to Save Energy, where he worked on energy efficiency policies and programs for eight years. He received a B.A. in economics from Amherst College, *magna cum laude*, and an M.B.A. from the Amos Tuck School of Business Administration at Dartmouth College, where he was elected an Edward Tuck Scholar.

David R. Wolcott recently joined RCG/Hagler Bailly, Inc. as manager of international demand-side management (DSM) programs. Under contract to the U.S. Agency for International Development and the World Bank, he is assisting Eastern European and Latin American countries to adopt utility integrated resource planning concepts. Previously, Mr. Wolcott directed the Integrated Resources Research Program at the New York State Energy Research and Development Authority. He designed and evaluated electric utility DSM programs, established roles for energy service companies through DSM bidding, modeled environmental externality costs, investigated regulatory incentive mechanisms, and analyzed natural gas DSM potential.

271

About the Authors

Charles J. Cicchetti is Managing Director of Arthur Andersen Economic Consulting in Los Angeles. He also teaches environmental economics at the University of Southern California in Los Angeles.

G. Alan Comnes received a B.S. in Science, Technology, and Society from Stanford University in 1984 and an M.A. in Energy and Resources from the University of California at Berkeley in 1987. He is a specialist in public utility resource planning and economics and currently works for the California Public Utilities Commission, Division of Ratepayer Advocates.

James E. Cuccaro is Manager of Economic Analysis at Orange and Rockland Utilities, Inc., an investor-owned electric and gas utility headquartered in Pearl River, NY. He is responsible for demand-side management planning and program evaluation, as well as load and energy forecasting, load research and economic and financial analysis.

Alan Destribats, at the time of this writing, was Vice President of Demand-Side Management and Least-Cost Planning for the New England Electric System where he oversaw the development, implementation, and evaluation of demand-side management programs and integrated resource planning for the NEES' subsidiaries operating in Massachusetts, New Hampshire, and Rhode Island. He is now a Senior Vice President with Synergic Resources Corporation.

Terry L. Dittrich is Director of Accounting at Orange and Rockland Utilities, Inc., an investor-owned electric and gas utility headquartered in Pearl River, N.Y. He is responsible for corporate accounting and taxation functions.

L. Mario DiValentino is Vice President, Accounting and Finance, and Controller of Orange and Rockland Utilities, Inc, an investor-owned electric and gas utility headquartered in Pearl River, N.Y. He was largely responsible for structuring his company's incentive rate-making mechanism.

Joseph Eto is a staff scientist in the Utility Planning and Policy Group of the Energy Analysis Program at the Lawrence Berkeley Laboratory where he works on the U.S. Department of Energy's Integrated Resources Planning Program. In 1988, he co-authored the National Association of Regulatory Utility Commissioner's *Handbook on Least-Cost Planning*, Volume 2, *The Demand-Side: Conceptual and Methodological Issues*.

Alan M. Freedman is Manager of Financial and Executive Communications at Orange and Rockland Utilities, Inc., an investor-owned electric and gas utility headquartered in Pearl River, N.Y., He is largely responsible for researching and writing financial and public policy issue papers, including the preparation of shareholder communications and testimony before regulatory agencies and legislative bodies.

Eric Hirst, at Oak Ridge National Laboratory since 1970, is a Corporate Fellow, a distinction shared by only 1% of the ORNL technical staff. His current interests include resource planning methods and results for electric utilities, with a focus on demand-side management programs.

Jennifer Jordan joined the staff of the American Council for an Energy-Efficient Economy in 1991. She works on utility, industry, and appliance issues. Prior to joining ACEEE, she worked on global warming and industrial energy issues at the Natural Resources Defense Council.

Steven Kihm has worked in the field of utility regulation for ten years, first at the Wisconsin Public Service Commission and later as a consultant with MSB Energy Associates, Inc. He is a nationally recognized expert in the areas of financial analysis, cost of capital, cost recovery methods, and utility management incentives. He has been a featured speaker at numerous national conferences.

Chris Marnay received a B.A. in Development Studies and an M.S. in Agricultural and Resource Economics from the University of California at Berkeley in 1981 and 1983, respectively, and is a Ph.D. candidate in the Energy and Resources Group. He is also a researcher for Lawrence Berkeley Laboratory and a consulting energy economist. His main professional interests are in public utility economics, electric utility resource planning, and the optimal dispatch of utility resources subject to environmental constraints.

Ellen K. Moran is Senior Consultant for Arthur Andersen Economic Consulting in Los Angeles. She specializes in economic and financial analysis of large-scale, capital-intensive projects, particularly those that are energy and utility-related.

David Moskovitz is a principal in the Regulatory Assistance Project, a non-profit agency that provides educational assistance in least-cost integrated resource planning to state public utility commissions. He served as a Commissioner on the Maine Public Utilities Commission from 1984 through 1989.

Paul Newman is the Assistant Administrator for Electric Policy with the Wisconsin PSC. He has been with the Wisconsin PSC for 15 years in a variety of roles, primarily involving the development and implementation of integrated resource planning and demand-side management programs for electric and gas utilities throughout the state. He is a member of the NARUC Staff Committee on Energy Conservation, a member of ASHRAE, and a Registered Professional Engineer in Wisconsin.

Richard Rosen is a senior scientist, a co-founder of Tellus Institute, and co-director of Tellus's energy group. Resources supply system modeling, economics, and pricing have been the major focus of Dr. Rosen's activities for the past 15 years at Tellus. Dr. Rosen received his Bachelor of Science degree from M.I.T. in 1966 and his Master's and Ph.D. degrees in physics from Columbia University in 1970 and 1974, respectively. Dr. Rosen's three-year appointed term on the Research Advisory Committee of the National Regulatory Research Institute was recently extended through June of 1992.

David Schoengold is a co-founder and principal of MSB Energy Associates, Inc. He has worked in the field of utility planning and regulation for 18 years, initially at the Wisconsin Public Service Commission and later as a consultant with MSB Energy Associates. He helped develop the principles of integrated least-cost resource planning while at the Wisconsin Commission. He is a nationally recognized expert in the areas of analytical methods and the application of those methods to utility planning.

Donald Schultz leads the Demand-Side Planning Section of the Division of Ratepayer Advocates, California Public Utilities Commission, which is responsible for regulatory review of utility demand-side management activities. Mr. Schultz was key participant in the California collaborative process, which gave rise to the shared-savings incentives described in this paper. In 1987, Mr. Schultz was co-project manager for California's revised *Standard Practice Manual for Economic Analysis of Demand-Side Management Programs.*

Gary B. Swofford is Vice President of Divisions and Customer Services at Puget Sound Power and Light Company, headquartered in Bellevue, Washington. He holds an electrical engineering degree from the University of Washington. Since 1980, he has directed the planning, design, and implementation of the company's energy conservation programs.

Index

accelerated incentive recovery, 109
Account Correcting for Efficiency
(ACE) mechanism, 35
accounting practices, and utility profitability, 6
ACE (Account Correcting for Efficiency) mechanism, 35
AER (Annual Energy Rate), 46
AFUDC (allowance for funds used
during construction) mechanism, 11, 34, 94
AJW (Averch-Johnson-Wellisz)
effect, ix–x, ixn
allowance for funds used during construction (AFUDC) mecha-
nism, 11, 34, 94
allowed rate of return (ROR), 84–
85
example, 87n
See also bonus returns
amortization periods, for ratebased
DSM expenditures, 85–86
*An Energy Efficiency Blueprint for
California* (California Collaborative Process), 29
Annual Energy Rate (AER), 46
annual rate relief, 203–4
Arizona, recent DSM incentive
developments (table), 32
attrition, 44–45, 56–57, 69
average rates
in Wisconsin and other midwest
states (chart), 207
in Wisconsin and similar fuel mix
states (chart), 208
Averch-Johnson-Wellisz (AJW)
effect, ix–x, ixn
avoided supply costs
increasing for the residential sector, 247
measuring, 106–7

balance accounts, 205
Bangor Hydro-electric Company
(BHE)
bid Payload Program, 174–77
dispute with the MPUC on DSM
program cost-effectiveness,
176–77
base costs, 65
correlations with sales and with
numbers of customers, 66,
67
proportion of variability in (R^2),
66n
base rate revenue requirement, 44
base rates, 39, 44, 80
effective, 40
base revenue growth, calculating, 69
BECo. *See* Boston Edison (BECo)
bench testing, 192
beta, 206n
BHE. *See* Bangor Hydro-electric
Company (BHE)
bill indexes, 141–45
common elements, 145
external, 142–43
internal, 143–45
See also bill indexing
bill indexing, 141–61, 263–64
administrative simplicity, 152–53
analysis (example), 193–94
attributes, 152–56
attributes (table), 262
as a comparative yardstick, 150
composite approaches
at CMP, 159–60
at NMPC, 157–59
conclusions, 160–61
cost control, 153
and the cost-effective use of electricity, 155
and cream-skimming, 153

and decoupling, 149–50, 257
designing incentive plan structures with, 145
determining utility accountability with, 143, 144
econometric approach at NMPC, 156–58
and environmental program costs, 155
evaluating LCP, 145–50, 160
external, 161
and externalities, 153n
and gaming, 153, 155, 160
internal, 161
long-run versus short-run efficacy, 149–50, 153–55, 161
non-participant impacts, 153
options, 142–45
    comparison by attributes (table), 156
    measuring bill reductions, 150–51
positive incentive effects, 147–48, 160
predictability, 152
purpose, 141
research results, 156–60
results at NMPC (table), 158
selecting customer control groups for, 144
selecting utility benchmark groups for, 142
suitability for LCP cost-effectiveness tests, 146
supply-side applications, 155
understandability, 152
See also customer bills
bill reductions
    annual (table), 154
    cumulative (table), 154
    measuring, 150–51
bills. See customer bills
bonus emissions allowances, 265
bonus returns, 22, 250, 258, 259, 263
    amounts authorized, 84–85
    attributes (table), 262

as bribes, 224
designing with bill indexes, 145
and DSM program cost control, 241, 250, 258
performance-based, 23–24, 85
    for less cost-effective measures, 15–16, 104, 110, 114, 114n
reports on, 241–42
and the shared-savings approach, 26–27
teamed with RDM, 30
See also shared-savings incentives
Boston Edison (BECo)
    DSM activity ratios (table), 237
    summary information on non-incentive DSM mechanisms (table), 233
bounty payments. See bonus returns
Bradford, Peter, ix–xi
bypass. See uneconomic bypass

California
    the California collaborative (on DSM), 29
    collaborative evaluation approach, 197
    concern about high incentive levels, 246
    development of ERAM, 46–52
    DSM program eligibility for shared-savings incentives, 113–14, 120n
        tables, 103, 114
    energy savings measurements, 246–47
    increase in DSM activity, 244–46
    large customer groups, 48–49, 51
    measured savings practices, 246–47
    NUGs, 51
    penalties for subpar performance, 29–30, 110
    performance criteria, 110, 116
    ratemaking system, 44–46, 127–28
    recent DSM incentive developments (table), 32

renewed interest in DSM, 51
shared-savings programs
  incentives calculation, 113
  proportion of all DSM activi-
    ties, 119
  shareholder earning caps, 112
  spending caps, 112
  tariffed rates, 45, 51
  See also California Public Utilities
    Commission; ERAM; Pacific
    Gas and Electric Company;
    Pacific Power and Light
    Company; San Diego Gas
    and Electric Company;
    Southern California Edison;
    Southern California Gas
California collaborative (on DSM),
  29, 192
  evaluation protocols, 264
California Public Utilities Commission (CPUC)
  adoption of ERAM, 39, 46–47
  AER suspension, 46
  decision to eliminate ERAM for
    large industrial customers,
    48–49
  and reversal, 49–50
  on DSM budget ratebasing, 83–
    84
  elimination of ERAM for PP&L,
    51–52, 52n, 179
  minimum DSM performance tar-
    gets, 179
  ratemaking procedures, 44–46
  reputation of, 55, 55n
  uneconomic bypass policy, 54
  utilities regulated by, 39n
  utility evaluation review, 197
  See also California
capital, cost of. *See* cost of capital
capital cost accounting, and supply-
  side cost recovery, 11
capitalization of DSM costs, 205
CenHud. *See* Central Hudson
  (CenHud)
CENTEL Electric, 249
  DSM activity ratios (table), 237

summary information on non-
  incentive DSM mechanisms
  (table), 233
Central Hudson (CenHud)
  DSM activity ratios (table), 236
  summary information on DSM
    incentives (table), 232
Central Maine Power (CMP)
  bill indexing results, 159–60,
    159n
  bonus formulas, 27
  proposal for shared-savings incen-
    tives on purchased power
    contracts, 266
  recent DSM developments, 33
  RPC decoupling success, 67
Chase, Bradford S., on DSM ratebasing, 81
Cicchetti, Charles J., 273
  on utility energy services pro-
    grams, 163–83
Cicchetti-Hogan proposal, 164
CL&P. *See* Connecticut Light and
  Power (CL&P)
Clean Air Act Amendments of 1990,
  bonus emissions allowances,
  265
CLF. *See* Conservation Law Foundation (CLF)
CMP. *See* Central Maine Power
  (CMP)
collaborative process
  in DSM planning and design, 16,
    28, 29, 116–17
  and evaluations, 197
Colorado, recent DSM incentive
  developments (table), 32. *See
  also* CENTEL Electric; Public Service of Colorado
ComElec. *See* Commonwealth Electric (ComElec)
commissions. *See* state regulatory
  commissions
Commonwealth Electric (ComElec)
  DSM activity ratios (table), 237
  inflated savings estimates, 241
  summary information on non-

incentive DSM mechanisms (table), 233
technique for calculating energy savings, 239n
use of DSM programs to reduce electric bills, 241
Comnes, G. Alan, 273
on California's ERAM experience, 39–59
company size, correlation with executive compensation, 222
competition
incentives for, 213–14
negative effects of ERAM on, 53–54, 55
competitive discounting, and rate-based DSM cost recovery, 92
competitiveness, and ratebasing DSM expenditures, 92, 93
ConEd. See Consolidated Edison (ConEd)
Connecticut
commission's DSM program encouragement, 240
decoupling implementation, 31
recent DSM incentive developments (table), 32
statute on DSM bonus returns, 85
See also United Illuminating
Connecticut Light and Power (CL&P)
attitude toward DSM, 241
DSM activity ratios (table), 237
summary information on non-incentive DSM mechanisms (table), 233
conservation. See energy conservation
conservation adder, 171
conservation and load management equipment, ratebasing, 82–84
conservation escrow accounts, 205
Conservation Law Foundation (CLF), work with NEES, 28, 197

Consolidated Edison (ConEd)
DSM activity ratios (table), 236
O&R type mechanism, 34
summary information on DSM incentives (table), 232
ConsPwr. See Consumers Power (ConsPwr)
Consumers Power (ConsPwr)
DSM activity ratios (table), 236
impacts of DSM incentives and penalties on, 249–50
recent DSM developments, 34
summary information on DSM incentives (table), 232
corporate culture, worship of growth, 221–22
cost attrition adjustments, 130–32
cost of capital
cost attrition adjustments for, 131–32
and ratebasing DSM expenditures, 81
and ratepayers' discount rates, 89–90
and risk redistribution, 73
cost recovery. See DSM cost recovery
cost-effectiveness calculation (PacifiCorp Energy Service Program), 171
cost-effectiveness tests, bill indexing and, 146
CPUC. See California Public Utilities Commission (CPUC)
cream skimming in DSM programs, 15–16, 104, 110, 246, 261
crossover point (between expensing and ratebasing), 89–90
Cuccaro, James E., 273
on RDM Plus incentives, 125–39
customer bills
annual and cumulative DSM impacts on, 153–54
comparing actual averages to LCP-predicted averages, 144
comparing overall averages to

control group averages, 144–45

comparing to average customer bills from other utilities, 142–43

comparing to econometric forecasts, 143

impacts of supply and demand-side actions on (table), 148

as a measure of total resource cost, 147–48

measuring reductions, 150–51

as reflections of actual DSM impacts, 148–49

as yardsticks in cost-effectiveness tests, 146

customer control groups, criteria for selecting, 144

customer costs, including/excluding in shared-savings programs, 107–8

customer service incentive (RDM Plus), 137–38

annual point allocation, 137

customers

impediments to making energy improvements, 170

supply and demand-side options (tables), 167, 168

See also large industrial customers

deadband (performance level), 110

decoupling (profits from sales), 9–10

bill indexing and, 149–50, 257

and energy savings measurement, 12–13

and fuel revenue accounting methods, 9

origins, 46–47

recent implementations, 31

and recoupling revenue to customer levels, 67

and risk redistribution, 55, 72–73

See also ERAM; RDM Plus; RDMs

Delmarva Power, Value Line Investment Survey on, 220

Demand-Side Management (DSM). See DSM (Demand-Side Management)

demand/supply options. See supply and demand-side options

Destribats, Alan, 273

on shared savings programs, 97–122

DetEd. See Detroit Edison (DetEd)

Detroit Edison (DetEd)

DSM activity ratios (table), 237

summary information on non-incentive DSM mechanisms (table), 233

Diablo Canyon nuclear plant, 23

discount rates (ratepayers')

expensing versus ratebasing, 90

and utilities' cost of capital, 89–90

discounting. See competitive discounting

disincentives for DSM programs, 10

focus on curing, 24–25

removing, 9–12, 52–53, 70

District of Columbia, recent DSM incentive developments (table), 32

Dittrich, Terry L., 273

on RDM Plus incentives, 125–39

DiValentino, L. Mario, 273

on RDM Plus incentives, 125–39

DSM (Demand-Side Management)

benefits of, 221

common utility view of, 215

and risk, 221

utilities and the development of, 163

See also DSM programs

DSM cost recovery, 11, 256

and bonus returns, 22, 23–24

energy service charge, 15

mechanisms providing, 256

the necessity of assurance, 225

risks of ratebasing, 91–92, 95

See also energy service charge; ratebasing DSM expenditures

DSM incentive mechanisms
  comparative assessment and future directions, 255–69
  DSM movement proposals for, 25
  removing disincentives, 9–12, 24–25, 52–53, 70
  reward/penalty mechanisms, 10, 33, 134–36
  summary of attributes (table), 262
  See also DSM incentives; DSM programs

DSM incentives
  alternatives to, 226
  amounts now being earned, 267
    table, 231–32
  caps on, 244, 246
  competition incentives, 213–14
  determining amounts, 266–67
    with evaluations and engineering estimates, 198, 199
    formulas for, 251
  for employees, 213, 247, 248
  federal interest in, 265
  for least-cost resources, 266
  as leveraged by evaluations (chart), 197
  limitations on, 145, 145n
  need for positive incentives, 11, 183
    presumptions regarding, 215–24, 229
  pegging to opportunity costs, 267
  and positive reinforcement, 222–24
  preliminary evaluation of, 229–51
  problems associated with, 212
  for publicly owned utilities, 265–66
  in RDM Plus, 133–37
  as start-up tools and occasional rewards, 224
  state regulatory commissions and, 203
  theories for belief in, 222–23

as unnecessary, 203–26, 267
See also bonus returns; DSM incentive mechanisms; DSM programs; positive incentives for DSM programs; preliminary evaluation of DSM incentives; regulatory incentives; shared-savings incentives

DSM investment
  factors affecting, 250
  and the rate base, 81, 93, 94
  regulatory incentives as impediments, 3–6
  RPC and, 73–74
    tables, 74, 75
  in the United States, vii, 2–3

DSM performance matrix, 134–36
  table, 135

DSM programs
  administrability, 14, 226, 259–60
  balance (fairness), 17, 226
  centralized operations, 119
  collaborative planning and design, 28
  collateral consequences, 260
  costs, 256
    capitalizing, 205
    determining, 198–99
    minimizing for consumers, 14, 260–61
  and cream skimming, 15–16, 104, 110, 246, 261
  current status, xi, 21
  current utility investments in, vii, 2–3
  early precedents, 21–24
  and effective regulatory reform, 7
  effectiveness, 268–69
  eligibility for shared-savings incentives, 113–14
  estimating energy savings from, 144–45, 190–95, 198–99
  evaluating, 7–8, 187–200
  evolution, 21–36
  flexibility, 196

at O&R, 30–31
and gaming, 16, 261
getting less cost-effective measures done, 15–16, 104, 110, 114, 114n
issues remaining, xi
measuring impacts, 12–13, 136
through bill indexing, 142–45, 150–51, 153–54
monitoring and review requirements, 226
non-participant impacts, 15, 166
ongoing questions, 269
as a paradox, 166
penalties in California, 29–30, 110
performance in California and New England, 117–21
table, 118
predictability, 13–14, 260
rate impacts, 268
reducing manipulation, 16
renewed interest in California, 51
risks, 98
scope, 17
selecting for evaluation, 199
slowing growth rates, 112n
spending caps, 112–13
trend-setting actions, 25–31
understandability, 13, 259
in the United States (table), 32–36
See also decoupling (profits from sales); DSM cost recovery; DSM incentive mechanisms; lost revenue recovery; shared-savings programs; and individual states and utility companies
DSM ratebasing. See ratebasing DSM expenditures

Eastern Utilities Associates (EUA)
DSM activity ratios (table), 237
inflated savings estimates, 241
summary information on non-

incentive DSM mechanisms (table), 233
technique for calculating energy savings, 239n
ECAC (Energy Cost Adjustment Clause) proceedings, 44, 46, 46n
econometric equations
for external bill indexing, 143
independent variables included at NMPC, 143n, 157
statistical performance in NMPC study, 157
effective base rates, 40. See also base rates
Electric Revenue Adjustment Mechanism. See ERAM (Electric Revenue Adjustment Mechanism)
electric utilities. See utility companies
electricity production, environmental effects, 2
electricity use estimates (household), table, 196
employee incentives, 213, 247, 248
end-use metering. See metering
energy commodities, and energy services, 164
energy conservation
fundamental problem for, x, 3
treatises on, x
See also energy efficiency
Energy Conservation Committee (NARUC), resolution on DSM and LCP, 25
Energy Cost Adjustment Clause (ECAC) proceedings, 44, 46, 46n
energy efficiency
beneficial effects, 2
New York State goal, 133
RDM Plus emphasis, 133
regulatory incentives as impediments, 3–6
regulatory reform and, 1

as a resource, 2–3
in the United States, 1–2
See also conservation; DSM
    programs
energy efficiency services. *See*
    energy services
energy efficient equipment, ratebas-
    ing, 82
energy prices, impacts of supply and
    demand-side actions on
    (table), 148
energy savings
    effects of evaluations on net bene-
        fits and utility incentive esti-
        mates (table), 195
    estimating for DSM programs,
        144–45, 190–95, 198–99
    impact of incentives on, 238–41,
        242
    charts, 240, 243
    RDM Plus determination strategy,
        136
energy service charge (ESC)
    as an antidote to overpayment and
        overinvestment, 169
    confusion regarding, 177–78
    as a disincentive for DSM pro-
        gram nonparticipants, 15,
        180–81
    for new homeowners and tenants,
        181
    in the PacifiCorp Energy Services
        Program, 171–73, 174–75,
        178–79
    the primary risk of, 15
    as protection for DSM program
        nonparticipants, 15, 166–
        68
    replacement for ERAM at PP&L,
        179
    the risk of disallowance, 178–79
energy service companies (ESCos),
    99
energy service tariff. *See* energy ser-
    vice charge (ESC)
energy services, 164–65, 264
    appearance of no real value, 180

See also energy service charge;
    energy services programs
energy services programs
    administrative complexity, 182
    advantages, 169–70
    attributes (table), 262
    and cross-subsidization, 183
    customer acceptance, 180, 182–
        83
    disadvantages, 180–82
    distinctive features, 182
    effectiveness of, 180–81
    facilitation of demand-side bid-
        ding, 169
    lack of utility incentive, 177–80,
        183
    OPUC on, 177
    and performance, 258–59
    and the total resource cost test,
        176–77
    as unbundling energy services,
        182
    and utility profits, 165
    See also Bangor Hydro-electric
        Company (BHE), bid Pay-
        load Program; energy service
        charge; PacifiCorp Electric
        Operations (PacifiCorp)
        Energy Services Program
engineering estimates, 190–92
    appropriate use of, 192, 260
    example, 193–94
    improving, 192, 198
    inaccuracy and inadequacy, 191
    and system design, 192
environmental damage costs, includ-
    ing in avoided cost measure-
    ments, 107
environmental effects of electricity
    production, 2
environmental groups, opposition to
    elimination of ERAM, 48
ERAM (Electric Revenue Adjust-
    ment Mechanism), x, 9, 23,
    39–59, 262–63
    administration, 57
    attributes (table), 262

and attrition, 45, 56–57
and authorized rates of return, 51n
the California collaborative and, 29
and competition, 53–54, 55
cooperative and contentious effects, 57, 58
development in California, 46–52
and DSM disincentives, 52–53
effects, 52–57
effort to eliminate, 48–50
elimination for PP&L, 51–52, 52n, 179
encouragement of innovative rate-making, 56
ERAM account, 40, 40n
ERAM rate, 40
and gaming, 56
growing interest in, 31
impact on promoting DSM programs, 127
lessons, 57–58
limited knowledge about, 58
mechanics, 40
mid-1980s review, 47–48
and NUGs, 51
opposition to, 50–51
origins, 23, 39, 46–47
projecting results to other states, 58–59
prudence reviews, 55
and rates, 50, 127
regulatory policy difficulties, 54, 58
and risk redistribution, 55
RPC compared to, 69–71
simplified example, 40–43
status of, 50–52
support for, 39–40, 56
and utility financial health, 54–55
ERAM account, 40, 40n
ERAM rate, 40
ESC. See energy service charge (ESC)
ESCos (energy service companies), 99

escrow accounting for DSM expenses, 205
estimating energy savings of DSM programs, 144–45, 190–92
iterative approach, 198
chart, 199
using multiple methods, 193–95, 198
See also evaluation(s) (of DSM programs)
Eto, Joseph, 273–74
on shared savings programs, 97–122
EUA. See Eastern Utilities Associates (EUA)
evaluation(s) (of DSM programs), 187–200, 260
applications, 188
by independent experts, 197–98
contracting out, 196
criteria for, 7–8
current efforts, 200
description, 188, 190, 196, 200
effects on net benefits and utility incentive estimates (table), 195
effects on PUC determination of utility incentives, 198
household electricity use estimates (table), 196
impact evaluations, 190
importance of, 264
leveraging of utility incentives, 195
chart, 197
preliminary evaluation of DSM incentives, 229–51
prior specification, 196
process evaluations, 189–90
reducing controversy in hearings, 195–98
resolving variations in estimates, 195–98
selecting programs for, 199
using the collaborative process, 197
See also estimating energy sav-

ings of DSM programs; preliminary evaluation of DSM incentives
expensing DSM expenditures, 80
externalities
and bill indexing, 153n
treatment of, 264–65

Federal Energy Regulatory Commission (FERC), regulation of wholesale rates, 115
FERC. *See* Federal Energy Regulatory Commission (FERC)
Financial Analysts Journal, on utility performance, 219
financial attrition, 45, 45n
financial incentives (for utility companies). *See* DSM incentives
firm size, correlation with executive compensation, 222
Florida, recent DSM incentive developments (table), 32
Ford Foundation Energy Policy Project, treatise on conservation, x
free drivers, 149n
free riders, 104, 149n
and the WEPCO incentive, 212
Freedman, Alan M., 274
on RDM Plus incentives, 125–39
front-loading payments, 109
FTY ratemaking. *See* future-test-year (FTY) ratemaking
fuel adjustment clauses
and decoupling, 9–10
and DSM cost recovery, 11
and utility profitability, x, 5–6
fuel cost adjustments. *See* fuel adjustment clauses
fuel offset mechanisms. *See* fuel adjustment clauses
fuel revenue accounting methods, and decoupling, 9
full reconciliations, for O&M cost attrition adjustments, 131
future-test-year (FTY) ratemaking, 64

advantages of, 204–5, 225
resistence to, 70

Gallagher, James, on incentive formulas, 251
gaming, 16, 261
and bill indexing, 153, 155, 160
ERAM effects, 56
RPC safeguards, 72
Geller, Howard S., vii–viii
general rate cases (GRCs) (in California), 44
mandated three-year cycle, 128
Georgia, recent DSM incentive developments (table), 32
Georgia Power, deferred DSM proposal, 32
goldplating, 261
Granite State Electric (GSE)
recent DSM developments, 34
shared-savings and overall DSM program performance, 117–21
table, 118
shared-savings programs
cost determination, 107–8
features (table), 103
incentives calculation, 113
maximizing incentive, 102, 104, 114n
performance thresholds, 104n, 111, 116
profitability, 119–20
summary information on DSM incentives (table), 231
GRCs. *See* general rate cases (GRCs) (in California)
Green Mountain Power (GMP), recent DSM developments, 35
growth (of utilities)
corporate culture and, 221–22
impact on returns, 215–19, 223
managerial incentives for, 221–22, 223
Wall Street's perspective, 219–21

GSE. *See* Granite State Electric (GSE)

Hawaii, recent DSM incentive developments (table), 32
hearings, reducing controversy over evaluations, 195–98
Hirst, Eric, 274
  on DSM and risk, 221
  on DSM program evaluation and financial incentives, 187–200
historic-test-year (HTY) ratemaking, 63–64
disadvantages of, 204
household electricity use estimates, table, 196
Howe, Jane Tripp, on above-average utility growth, 220–21
HTY ratemaking. *See* historic-test-year (HTY) ratemaking
"Hypothetical Ideal Case" (PacifiCorp Energy Services Program), 172–73
  table, 174–75

Idaho
  DSM ratebasing decision, 82–83
  recent DSM incentive developments (table), 33
Illinois, recent DSM incentive developments (table), 33
impact evaluations, 190
incentives. *See* managerial incentives; perverse incentives; positive incentives for DSM programs; regulatory incentives; shared-savings incentives; tax incentives
independent power, incentive possibilities, 266
indexing. *See* bill indexing
Indiana, recent DSM incentive developments (table), 33
inflation adjustments, for O&M cost attrition adjustmemnts, 131
information-only programs, 199

integrated resource planning. *See* least-cost planning (LCP)
Iowa
  bonus formulas, 27
  recent DSM incentive developments (table), 33
IRP (integrated resource planning). *See* least-cost planning (LCP)
iterative approach for estimating energy savings, 198
  chart, 199

JCP&L. *See* Jersey Central Power and Light (JCP&L)
Jersey Central Power and Light (JCP&L)
  DSM activity ratios (table), 237
  ratebased DSM amortization, 86
  summary information on non-incentive DSM mechanisms (table), 233
Jones, Robert, on utility stock performance, 219
Jordan, Jennifer A., 274
  preliminary evaluation of DSM incentives, 229–51

K factor, 69
Kansas, statute on DSM bonus returns, 84–85
Kilm, Steven, 274
  on DSM incentives as unnecessary, 203–26

large industrial customers
  attitudes toward RDMs, 48, 51, 52
  separation from other rate classes in California, 48–49
Large Light and Power (LL&P) customer class, separation from other rate classes in California, 48–49
LCP. *See* least-cost planning (LCP)
least-cost planning (LCP), 2–3, 24–25

aligning financial incentives with, 8–9
evaluating through bill indexing, 145–50, 160
including supply-side aspects, 17
the major obstacle, 3
"leveling the playing field," and ratebasing DSM expenditures, 81
LILCo. *See* Long Island Lighting Company (LILCo)
LL&P (Large Light and Power) customer class, separation from other rate classes in California, 48–49
load management equipment ratebasing, 82
load reductions, measuring, 104–6. *See also* estimating energy savings of DSM programs
load-control equipment ratebasing, 82
loans (DSM), ratebasing, 82, 83
Long Island Lighting Company (LILCo)
  DSM activity ratios (table), 236
  lost revenue recovery, 26n, 243–44
  summary information on DSM incentives (table), 232
lost revenue recovery, 10–11, 26, 26n, 243–44, 256–57
  compared with decoupling, 70n
  mechanisms providing, 257
  in revised rates, 28, 28n
  through FTY ratemaking, 204
  See also DSM cost recovery
lost revenue treatment, 26, 114–15, 115n
Lovins, Amory, treatise on conservation, x

Madison Gas and Electric Company (MG&E), competition incentive, 213–14
Maine

base revenue growth calculation, 69
decoupling implementation, 31
DSM ratebasing statute, 83
fuel revenue accounting methods, 9
recent DSM incentive developments (table), 33
See also Central Maine Power (CMP)
Maine Public Utilities Commission (MPUC)
  dispute with BHE on DSM program cost-effectiveness, 176–77
  DSM cost recovery and incentive policy, 179–80, 180
  on savings evaluations, 200
managerial incentives, for growth, 221–22, 223
marketing, disadvantages of discouraging, 53
markups, 259, 263
  attributes (table), 262
Marnay, Chris, 274
  on California's ERAM experience, 39–59
Maryland, recent DSM incentive developments (table), 33
Massachusetts
  collaborative evaluation approach, 197
  commission's DSM program encouragement, 240
  DSM ratebasing decision, 83
  NEES DSM program, 27–28
  recent DSM incentive developments (table), 33
  See also Boston Edison; Commonwealth Electric; Eastern Utilities Associates; Massachusetts Electric; Western Massachusetts Electric Company
Massachusetts Electric (ME), 101n
  measured savings shortfalls, 242

recent DSM developments, 33
summary information on DSM
    incentives (table), 231
maximizing incentives, including in
    shared-savings programs,
    102, 104, 114n
ME. *See* Massachusetts Electric
    (ME)
measure funding limit (PacifiCorp
    Energy Service Program),
    171
MetEd. *See* Metropolitan Edison
    (MetEd)
metering, 192
    analysis (example), 193–94
Metropolitan Edison (MetEd)
    DSM activity ratios (table), 237
    summary information on non-
        incentive DSM mechanisms
        (table), 233
MG&E (Madison Gas and Electric
    Company), competition
    incentive, 213–14
Michigan, recent DSM incentive
    developments (table), 34. *See
    also* Consumers Power;
    Detroit Edison
Michigan Public Service Commis-
    sion, DSM authorizations for
    Consumers Power, 249
minimum performance requirements
    for utility companies, 187–
    88
Minnesota, recent DSM incentive
    developments (table), 34
Model Conservation Standards
    estimates of effects, 195
    table, 196
Montana, statute on DSM bonus
    returns, 85
Moran, Ellen K., 274
    on utility energy services pro-
        grams, 163–83
Moskovitz, David, 275
    on bill indexing, 141–61
    on engineering estimates, 191

on fuel revenue accounting, 9
on regulation, LCP, and DSM,
    24–25
on regulatory reform for DSM, 1–
    18
on RPC decoupling, 63–76
MPUC. *See* Maine Public Utilities
    Commission (MPUC)
multi-staged filings, 130

Nadel, Steven M., 271
    on DSM incentive mechanisms,
        255–69
    preliminary evaluation of DSM
        incentives, 229–51
Narragansett Electric (NE)
    recent DSM developments, 35
    shared-savings and overall DSM
        program performance, 117–
        21
    table, 118
    shared-savings programs
        cost determination, 107–8
        features (table), 103
        incentives calculation, 104,
            105, 113
        maximizing incentive, 102,
            104, 114n
        performance thresholds, 104n,
            111, 116
        profitability, 119–20
    summary information on DSM
        incentives (table), 231
NARUC (National Association of
    Regulatory Utility Commis-
    sioners), resolution on regu-
    latory reform, 6–7
NARUC Energy Conservation Com-
    mittee, resolution on DSM
    and LCP, 25
National Association of Regulatory
    Utility Commissioners
    (NARUC), resolution on reg-
    ulatory reform, 6–7
National Energy Strategy, on energy
    efficient investment, 2

Natural Resources Defense Council
(NRDC)
and the California collaborative,
29
opposition to elimination of
ERAM, 48
NE. *See* Narragansett Electric (NE)
NEES. *See* New England Electric
System (NEES)
net benefits. *See* net savings
net resource value, formula for calculating, 100, 133
net revenue losses, OPUC on, 178
net savings, 190
estimating, 194
table, 195
Nevada, recent DSM incentive
developments (table), 34. *See
also* Nevada Power; Sierra
Pacific
Nevada Power (NevPwr)
DSM activity ratios (table), 237
summary information on non-
incentive DSM mechanisms
(table), 233
NevPwr. *See* Nevada Power
(NevPwr)
New England
advantages of DSM resources,
240–41
current utility investments in
DSM programs, 2–3
impact of incentives on DSM
activity, 238–41
chart, 240
NEES DSM programs, 27–28
reports on bonus returns, 241–42
utility commissions' DSM program encouragement, 240
See also New England Electric
System; United Illuminating;
Western Massachusetts
Electric
New England Electric System
(NEES)
calculating earnings from shared-

savings mechanisms, 100,
100n, 102, 104
DSM activity ratios (table), 236
on DSM and risk, 221
DSM program eligibility for
shared-savings incentives,
113
table, 103
DSM program evaluation efforts,
106, 106n
DSM program performance, 119
criteria for, 110–11, 110n
table, 118
evaluation
commitments to, 200
iterative approach, 198
extensive DSM efforts, 241, 241n
impact of incentives on attitude
toward DSM, 241
multi-state DSM programs, 27–28
RPC decoupling success, 67
summary information on DSM
incentives (table), 231
technique for calculating energy
savings, 239n
wholesale subsidiary's revenue
adjustment mechanism, 115
See also Granite State Electric;
Massachusetts Electric; Narragansett Electric
New England Power, revenue adjustment mechanism, 115
New Hampshire
NEES DSM program, 27–28
recent DSM incentive developments (table), 34
See also Granite State Electric
New Jersey
decline in DSM activity, 242n
impact of incentives on DSM
activity, chart, 243
recent DSM incentive developments (table), 34
See also Jersey Central Power and
Light; Public Service Electric
and Gas

New York Public Service
  Commission
  opinion on ratemaking and DSM,
    26, 125–26
  penalties for subpar performance,
    244
New York State
  adoption of RDMs, 52
  energy efficiency goal, 133
  evaluation approach, 198
  impact of incentives on DSM
    activity, 242
  chart, 243
  incentive mechanisms in effect,
    243
  increase in DSM activity, 242–43
  industrial customer opposition to
    RDMs, 51, 52
  lost revenue recovery, 26, 26n
  multi-staged filings, 130
  O&R incentive revision, 30–31
  O&R/NIMO decision, 26–27
  recent DSM incentive develop-
    ments (table), 34
  shareholder earning caps, 112–13
  See also Central Hudson; Consoli-
    dated Edison; Long Island
    Lighting Company; New
    York State Electric and Gas;
    Niagara Mohawk Power Cor-
    poration; Orange and Rock-
    land Utilities; Rochester Gas
    and Electric Company
New York State Electric and Gas
  (NYSEG)
  DSM activity ratios (table), 236
  summary information on DSM
    incentives (table), 232
Newman, Paul, 275
  on DSM incentives as unneces-
    sary, 203–26
Newport Electric, recent DSM
    developments, 35
Niagara Mohawk Power Corporation
  (NMPC)
  bill indexing results, 156–59

table, 158
DSM activity ratios (table), 236
O&R/NIMO decision, 26–27
recent DSM developments, 34
simulated annealing technique,
  142n, 157
summary information on DSM
  incentives (table), 232
NIMO/O&R decision. See O&R/
  NIMO (Orange and Rock-
  land/Niagara Mohawk)
  decision
NMPC. See Niagara Mohawk Power
  Corporation (NMPC)
nonprogram savings, 190
nonutility generators (NUGs), in
  California, 51
North Carolina, recent DSM incen-
  tive developments (table), 34
Northern States Power (Minnesota),
  recent DSM developments,
  34
NRDC. See Natural Resources
  Defense Council (NRDC)
NUGs (nonutility generators), in
  California, 51
NYSEG. See New York State Elec-
  tric and Gas (NYSEG)

O&R. See Orange and Rockland
  Utilities (O&R)
O&R/NIMO (Orange and Rockland/
  Niagara Mohawk) decision,
  26–27
Ohio, recent DSM incentive developments (table), 35
operation and maintenance (O&M)
  expenses, cost attrition
  adjustments for, 130–31
operational attrition, 45
OPUC. See Oregon Public Utilities
  Commission (OPUC)
Orange and Rockland Utilities
  (O&R)
  on alternative ratemaking models,
  126–28

concerns about existing ratemaking practices, 126
DSM activity ratios (table), 236
flexibility of DSM programs, 30–31
O&R incentive revision, 30–31, 34
O&R/NIMO decision, 26–27
peak load reduction and energy efficiency goals (table), 134
ratemaking goals, 128–29
RDM proposal, 129
summary information on DSM incentives (table), 232
views on ERAM, 127–28
See also RDM Plus (revenue-decoupling-plus-incentives mechanism)
Orange and Rockland/Niagara Mohawk (O&R/NIMO) decision, 26–27
Oregon
bonus formulas, 27
DSM loan ratebasing, 82
measured savings practices, 247
recent DSM incentive developments (table), 35
See also Pacific Power and Light Company; Portland General Electric
Oregon Public Utilities Commission (OPUC)
confusion regarding ESC, 177–78
on the PacifiCorp Energy Services Program, 177
on the treatment of net revenue losses, 178

Pacific Gas and Electric Company (PG&E)
alleged cream skimming, 246
argument for retaining ERAM, 50
DSM activity ratios (table), 236
DSM program eligibility for shared-savings incentives, 113–14, 120n

table, 114
DSM program evaluation efforts, 106
ERAM proposal, 23, 46–47
performance thresholds (table), 111
revenue, sales, and customer totals (table), 102
shared-savings and overall DSM program performance, 117–21
table, 118
shared-savings programs, 32
cost and incentive determination, 107–8, 122, 247
features (table), 103
profitability, 119–20, 120n
summary information on DSM incentives (table), 231
Pacific Northwest, current utility investments in DSM programs, 2–3
Pacific Northwest Electric Power Planning and Conservation Act of 1990, 171
Pacific Power and Light Company (PP&L)
authorized incentives, 244n
DSM activity ratios (table), 237
elimination of ERAM, 51–52, 52n, 179
energy services proposal, 33, 35
low-flow showerhead metering experiment, 192
ratebased weatherization programs, 82
summary information on non-incentive DSM mechanisms (table), 233
See also PacifiCorp Electric Operations (PacifiCorp) Energy Services Program
PacifiCorp Electric Operations (PacifiCorp) Energy Services Program, 170–74
administrative complexity, 182

in California
  accounting treatment for DSM
    program expenditures, 179
  first year incentive, 179
  three-tiered DSM cost recovery
    approach, 179
in Oregon
  availability of 100% financing,
    173
  conservation adder, 171
  cost-effectiveness calculation
    (measure funding limit), 171
  energy conservation measures,
    171
  energy service charge, 171–73,
    174–75, 178–79
  energy services contract protec-
    tion measures, 181
  general accounting treatment
    for conservation-related
    expenses, 178
  proposed assessment, 173
  sharing of program benefits,
    172–73
  previous DSM programs, 170–71
  See also energy services programs
PAP&L. *See* Pennsylvania Power
  and Light (PAP&L)
participation, as a measure of program performance, 110
penalties for subpar performance
  in California, 29–30, 110
  in New York, 244
Pennsylvania
  decline in DSM activity, 242n
  impact of incentives on DSM
    activity, chart, 243
  recent DSM incentive develop-
    ments (table), 35
  See also Metropolitan Edison;
    Pennsylvania Power and
    Light
Pennsylvania Power and Light
  (PAP&L)
  DSM activity ratios (table), 237
  summary information on non-

incentive DSM mechanisms
  (table), 233
PEPCO. *See* Potomac Electric Power
  Company (PEPCO)
performance (of DSM programs)
  basing bonus returns on, 23–24,
    85
  and measurement, 259
  mechanisms rewarding, 258
  need for detailed information on,
    189
  the RDM Plus DSM performance
    matrix, 134–36
    table, 135
  See also bonus returns
performance requirements for utility
  companies, minimum, 187–
  88
performance thresholds, 104n, 109–
  11, 116
performance-based earnings adders,
  114
perverse incentives, of engineering
  estimates, 191–92
PG&E. *See* Pacific Gas and Electric
  Company (PG&E)
PGE. *See* Portland General Electric
  (PGE)
Portland General Electric (PGE)
  bonus formulas, 27
  DSM activity ratios (table), 236
  impact of the DSM incentive, 246
  increased residential avoided cost
    value, 247–48
  recent DSM developments, 35
  summary information on DSM
    incentives (table), 231
positive incentives for DSM
  programs
  basic characteristics, 132
  caps, 258
  factors determining, 258
  mechanisms providing, 257–58
  need for, 11, 183, 257
  presumptions regarding, 215–24,
    229

RPC effects, 70–71
thresholds, 258
See also bonus returns; shared-savings incentives
positive reinforcement, and DSM incentives, 222–24
Potomac Electric Power Company (PEPCO)
recent DSM developments, 32, 33
RPC decoupling success, 67
power
marginal cost to utility companies, 5
purchased power clauses, 6
PP&L. *See* Pacific Power and Light Company (PP&L)
preliminary evaluation of DSM incentives, 229–51
approach, 230–34
comparisons of incentive mechanisms, 238
conclusions, 250–51
effects of amounts on DSM activity, 238
effects on DSM activity, chart, 239
median increases in DSM activity, 235
exceptions to the pattern, 238
net savings calculations, 234
overall analysis, 235–38
pre- and post-incentive year determination, 233n–34n
rank-sum statistical test, 235–38
ratios calculated, 233–34
ratios reported, 235
and compared, 235, 238
table, 236–37
regional analysis, 238–50
of Colorado utilities, 249
of Michigan utilities, 249–50
of New England utilities, 238–42
of New York State utilities, 242–44
of west coast utilities, 244–48
of Wisconsin utilities, 248–49

summary information for control group utilities (table), 233
summary information for utilities receiving incentives (table), 231–32
price signals, and ratebasing DSM expenditures, 81
prices. *See* energy prices
prior authorization, the right of, 225
process evaluations, 189–90
profits, decoupling from sales. *See* decoupling (profits from sales)
Profits and Progress Through Least-Cost Planning (Moskovitz), 64
program evaluations. *See* evaluation(s) (of DSM programs)
prudence reviews, and ERAM, 55
PSCO. *See* Public Service of Colorado (PSCO)
PSE&G. *See* Public Service Electric and Gas (PSE&G)
Public Service Electric and Gas (PSE&G)
DSM activity ratios (table), 237
summary information on non-incentive DSM mechanisms (table), 233
Public Service of Colorado (PSCO)
DSM activity ratios (table), 236
DSM programs, 32
impact of DSM incentive, 249
RPC decoupling success, 67
summary information on DSM incentives (table), 231
Public Utility Regulatory Policies Act, adding DSM incentives, 265
Puget Power. *See* Puget Sound Power and Light Company (Puget Power)
Puget Sound Power and Light Company (Puget Power)
development of RPC, 64–68
evaluation plan, 200
impacts of RPC, 73–75

ratebased DSM budget, 22, 83, 86
recent DSM developments, 35
RPC decoupling success, 67
    regression analyses, 65–66, 67
stock price increases, 73–74
Value Line Investment Survey on,
    74
See also RPC (revenue-per-
    customer decoupling
    mechanism)
purchased power clauses
    and DSM cost recovery, 11
    RPC decoupling and, 65, 65n
    and utility profitability, 6

R^2 value, 66n
rank-sum statistical test, 235, 238
Rappaport, Alfred, on firm size and
    executive compensation, 222
rate base, 79
    and DSM investment, 81, 93, 94
rate base investment, cost attrition
    adjustments for, 131
rate relief. *See* annual rate relief
rate-of-return bonuses. *See* bonus
    returns
ratebasing. *See* ratebasing DSM
    expenditures
ratebasing DSM expenditures, 9, 22,
    63–76, 79–80, 79–96, 263
    allowed rate of return, 84–85
    amortization periods, 85–86
    attributes (table), 262
    balance sheet risks, 92–93
    benefits, 93
    and competitiveness, 92, 93
    cost recovery risks, 91–92
    as a cost-recovery mechanism,
        94–95
    description, 80
    as a DSM incentive mechanism,
        95–96
    and DSM status, 93
    eligible expenditures, 82–84
    equity, 82
    expansive, 83–84

or expensing DSM expenditures,
        86–90
    tables, 87, 88, 89
    financial incentives, 81
    "leveling the playing field," 81
    and price signals, 81
    and the rate base, 81, 93, 94
    and rate stability, 82, 93
    and ratepayers' discount rates,
        89–90
    reasons for, 80–82
    and revenue requirements, 86–90
    tables, 87, 88, 89
    risks, 91–93
    and utility cash flow, 90–91
    See also bonus returns
ratemaking
    alternative models, 126–28
    in California, 44–46, 51, 56
    and incentives, ix, 3–6
    and utility profitability, 4–5
    See also future-test-year (FTY)
        ratemaking; historic-test-year
        (HTY) ratemaking; regula-
        tion; regulatory reform(s)
ratepayers' discount rates. *See* dis-
    count rates (ratepayers')
rates. *See* average rates; relative rate
    increases
RDM Plus (revenue-decoupling-
    plus-incentives mechanism),
    125–39
    benefits and applicability, 138–39
    broad-based objectives, 132
    complete concept, 129, 130
    comprehensive nature, 138
    cost attrition adjustments, 130–32
    customer service incentive, 137–
        38
    development, 128–32
    DSM incentive
        goals, 133
        performance matrix structure,
            134–36
        reward/penalty range, 133
        strengths and possible improve-
            ments, 137

DSM programs, 133–34
  determining impacts, 136
  implications for DSM, 136–37
  net resource savings (definition),
    133
  positive incentives, 132
  savings determination strategy,
    136
  treatment of revenues, 129–30
RDMs (revenue decoupling mecha-
    nisms), 257
  in the natural gas industry, 46,
    46n
  for O&R, 30–31
  opposition to, 51, 52
  outside California, 52
  See also ERAM; RDM Plus;
    RPC
reconciliations. *See* full reconcilia-
    tions; volumetric
    reconciliations
reform. *See* regulatory reform(s)
regression analyses, 66n
  for Central Maine Power (table),
    67
  for Puget Power, 65–66, 67
regulation
  and changing environments, 91–
    92
  conventional, 1, 3
    and DSM, 24–25
  fundamental change in, 269
  guidelines in, 13–14
  making the traditional approach
    work for DSM, 225, 226
  the right of prior authorization,
    225
  risks of nontraditional methods,
    98
  test of reason, 65
  wise and unsound, ix
  See also ratemaking; regulatory
    incentives; regulatory instru-
    ments; regulatory reform(s)
regulatory commissions. *See* state
    regulatory commissions
regulatory incentives, ix, 187

  aligning with LCP, 8–9
  benefitting shareholders, 24
  development of, 269
  for DSM programs, vii
  the fundamental problem with, x,
    3
  as impediments to energy effi-
    ciency investment, 3–6
  proper rewards, 127
  for publicly owned utilities, 265–
    66
  See also disincentives for DSM
    programs; DSM incentives;
    perverse incentives; positive
    incentives for DSM pro-
    grams; shared-savings
    incentives
*Regulatory Incentives for Demand-
    Side Management*, overview,
    vii–viii, 18
regulatory instruments, and utility
    profitability, 5–6
regulatory philosophy, importance
    of, 206, 225
regulatory reform(s)
  criteria for evaluating, 7–8
  current status, xi, 21
  and DSM programs, 7
  and energy efficiency, 1
  NARUC on, 6–7
  See also decoupling (profits from
    sales); DSM programs
Reid, Michael, 271
  on the evolution of DSM incentive
    programs, 21–36
  on ratebasing DSM expenditures,
    79–96
relative rate increases
  in Wisconsin and other midwest
    states (chart), 209
  in Wisconsin and similar fuel mix
    states (chart), 210
resource costs, 65
return on equity (ROE) adjustments,
    30
  attributes (table), 262
revenue adjustment mechanisms. *See*

RDMs (revenue decoupling mechanisms)
revenue decoupling mechanisms. *See* RDMs (revenue decoupling mechanisms)
revenue growth. *See* base revenue growth
revenue-per-customer decoupling. *See* RPC (revenue-per-customer decoupling mechanism)
revenues, treatment in RDM Plus, 129–30
revised residential end-use forecast (WEPCO), 212
reward/penalty mechanisms, 33
  as decoupling measures, 10
  the RDM Plus DSM performance matrix, 134–36
    table, 135
  See also penalties
Rhode Island
  NEES DSM program, 27–28
  recent DSM incentive developments (table), 35
  See also Eastern Utilities Associates; Narragansett Electric
risk redistribution
  collaborative process agreements, 117
  under ERAM, 55
  under RPC, 72–73
  and utility cost of capital, 73
Rochester Gas and Electric Company
  DSM activity ratios (table), 236
  summary information on DSM incentives (table), 232
ROE (return on equity) adjustments, 30, 258, 263
ROR. *See* allowed rate of return (ROR)
Rosen, Richard, 275
  on bill indexing, 141–61
Rowe, John, on NEES DSM success, 28
RPC (revenue-per-customer decoupling mechanism), 9, 63–76, 263
  allowed revenue, 70
  applicability, 64
  assessment, 75–76
  attributes (table), 262
  calculating, 68
  class-specific, 71
  compared to ERAM, 69–71
  conclusions of studies, 67
  customer size and mix considerations, 71–72
  description, 64–65
  development at Puget Power, 64–68
  and DSM disincentives, 70
  and DSM investment at Puget Power, 73
    table, 74
  and DSM savings at Puget Power, 73
    table, 75
  fine tuning, 68–69
  gaming safeguards, 72
  impacts on Puget Power, 73–75
  incentive to attract new customers, 72
  and investor response to Puget Power, 73–74
    table, 76
  mechanics, 68
  positive DSM incentive effects, 70–71
  potential problems, 71–73
  and purchased power clauses, 65, 65n
  and risk redistribution, 72–73

sales
  and corporate culture, 222
  decoupling profits from. *See* decoupling (profits from sales)
  and managers' salaries, 222
  See also growth
San Diego Gas and Electric Company (SDG&E)

DSM activity ratios (table), 236
DSM incentive success, 246
DSM program eligibility for shared-savings incentives, 113–14, 120n
DSM program evaluation efforts, 106
employee incentive, 247
revenue, sales, and customer totals (table), 102
shared-savings and overall DSM program performance, 117–21
shared-savings programs, 32
    cost and incentive determination, 107–8, 247
    and cost control, 247, 247n
    features (table), 103
    profitability, 119–20
summary information on DSM incentives (table), 231
SCE. See Southern California Edison (SCE)
SCG. See Southern California Gas (SCG)
Schlien, Milton, on growth and earnings at Delmarpa Power, 220
Schoengold, David, 275
    on DSM incentives as unnecessary, 203–26
Schultz, Donald, 275
    on shared savings programs, 97–122
SDG&E. See San Diego Gas and Electric Company (SDG&E)
shared-savings agreements, 99
shared-savings incentives
    basing on utility or total resource costs, 107–8, 247
    as components of DSM programs, 118–19
    failures, 215
    including maximizing incentives with, 102, 104, 114n
    and program cost control, 247
    program eligibility for, 113–14

on purchased power contracts, 266
recovering, 109
typical, 104
See also bonus returns
shared-savings mechanisms, 258, 263
    attributes (table), 262
    basic idea, 97
    designing with bill indexes, 145
    encouragement of large customer programs, 241–42, 243, 250
    evaluation ratios, 117–18
    and future year estimates, 106n
    popularity, 193
    See also shared-savings programs
shared-savings programs, 26–27, 97–122, 259–60
    avoided cost measurements, 106–7
    collaborative design of, 116–17
    comparisons among, 115–16
    complexity of, 250
    costs
        definitions, 100, 100n
        measuring, 107–8
        recovering, 108
        to society, 120–21, 120n
    and differences among utilities, 101–2
    earnings (benefits)
        calculating, 97–98, 100, 100n, 187
        chart, 189
        utilities' shares, 100, 100n, 102, 104, 189, 194–95
    evaluations in progress, 106
    as experiments, 100–101
    features, 101, 102
        table, 103
    load reductions measurement, 104–6
    and lost revenues, 114–15
    origin of, 99
    performance in California and New England, 117–21
        table, 118

performance thresholds, 104n, 109–11, 116
performance-based rewards, 100, 114
profitability, 119–20, 119n
proportion of all DSM activities, 119, 119n
the regulator as arbiter, 99
risks of, 98, 106
shareholder earning caps, 112–13, 244
spending caps, 112
summary, 121–22
utilities' earnings from, 115
See also Granite State Electric; Narragansett Electric; Pacific Gas and Electric Company; San Diego Gas and Electric Company; shared-savings incentives; shared-savings mechanisms
shareholder earning caps in shared-savings DSM programs, 112–13
shareholder incentives. *See* bonus returns
Sierra Pacific (SP)
DSM activity ratios (table), 237
increase in DSM activity, 245n
summary information on non-incentive DSM mechanisms (table), 233
simulated annealing technique, 142n, 157
Southern California Edison (SCE)
DSM activity ratios (table), 236
ratebased DSM program, 32, 83–84
summary information on DSM incentives (table), 231
Southern California Gas (SCG), recent DSM developments, 32
SP. *See* Sierra Pacific (SP)
spending caps on shared-savings DSM programs, 112–13
state regulatory commissions

decision-making responsibility, 209–11
and DSM incentives, 203
on LCP, 2
proactive involvement of, 209–11, 225, 242
regulatory philosophy and, 206, 225
Sterzinger, George, on developing DSM programs, 226
stock betas, 206n
stock prices (of utilities)
increases at Puget Power, 73–74
versus construction expenditures (chart), 220
Wall Street's perspective, 219–21
supply and demand-side options
for customers (tables), 167, 168
for utility companies (tables), 165, 167, 168
Swofford, Gary B., 276
on RPC decoupling, 63–76

T&D (transmission and distribution) costs, including in avoided cost measurements, 107
tariffed rates in California, 45, 51
tax incentives (of the early 1970s), and utility energy conservation, x
test of reason, 65
Tonn, B., estimates of Model Conservation Standards effects, 195
total resource cost, customer bills as a measure of, 147–48
total resource cost test (TRC), 146
conflicts with energy services programs, 176–77
total savings, definition, 190
transmission and distribution (T&D) costs, including in avoided cost measurements, 107

UI. *See* United Illuminating (UI)
uneconomic bypass, efforts in preventing, 47, 54

United Illuminating (UI)
 DSM activity ratios (table), 236
 extensive DSM efforts, 241, 241n
 impact of incentives on attitude
  toward DSM, 241
 recent DSM developments, 32
 small DSM incentive, 241
 summary information on DSM
  incentives (table), 231
United States
 current utility investments in
  DSM programs, vii, 2–3
 energy efficiency, 1–2
 federal energy policy, xi
 recent DSM incentive developments (table), 32–36
utility companies
 aligning financial incentives with
  LCP, 8–9
 annual returns to stockholders
  (chart), 217
 annual sales growth for slow- and
  fast-growing utilities (chart),
  218
 annual sales growth rates (chart),
  216
 annual total returns for slow- and
  fast-growing utilities (chart),
  219
 calculating earnings from shared-
  savings mechanisms, 100,
  100n, 102, 104, 194–95
  table, 195
 common view of DSM, 215
 comparison of (table), 102
 current investments in DSM programs, vii, 2–3
 customer bills as a measure of
  performance, 147–48
 decoupling profits from sales, 9–
  10
 determining accountability in bill
  indexing, 143, 144
 and the development of DSM,
  163
 DSM ratebasing and cash flow,
  90–91

electric utility stocks versus gas
  utility stocks, 217–18
 financial health under ERAM,
  54–55
 and growth
  impact on returns, 215–19, 223
  Wall Street's perspective, 219–
  21
 the impact of investment returns
  on market value, 205–6,
  206n
 incentives to maximize sales, 3–4
 mimimum performance requirements, 187–88
 minimizing prices and/or revenue
  requirements, 146
 ratemaking and profitability, 4–5
 selecting benchmark groups for
  bill indexing, 142
 selecting customer control groups
  for bill indexing, 144
 shared-savings program earnings,
  115
 supply and demand-side options
  (tables), 165, 167, 168
 supply options as a source of
  earnings, 222–23
 traditional investment focus, ix–x,
  3
 See also cost of capital; stock
  prices; utility managers; and
  individual utility companies
utility customers. See customers
utility managers
 confusion over DSM risks, 221
 preoccupation with sales, 221–22
 primary salary determinants, 222

Value Line Investment Survey
 on Delmarva Power, 220
 on Puget Power, 74
Vermont, recent DSM incentive
  developments (table), 35
Vermont Public Service Board, on
  DSM and risk, 221
Virginia, recent DSM incentive
  developments (table), 35

volumetric reconciliations, for O&M
cost attrition adjustments,
131

Walmet, Gunnar E., vii–viii
Washington State
base revenue growth calculation,
69
decoupling implementation, 31
DSM incentives legislation, 21–22
early DSM program, 9, 21–22
recent DSM incentive developments (table), 35
statute on DSM bonus returns, 84
See also Puget Sound Power and
Light Company
weatherization programs, 82, 84
WEPCO. *See* Wisconsin Electric
Power Company (WEPCO)
Western Massachusetts Electric
Company (WMECO)
impact of incentives on attitude
toward DSM, 241
recent DSM developments, 33
summary information on DSM
incentives (table), 231
White, D., estimates of Model Conservation Standards effects,
195
Wisconsin
annual rate relief, 203–4
average rates (charts), 207, 208
current utility investments in
DSM programs, 2–3
experience with DSM incentives,
211–15
recent DSM incentive developments (table), 36
regulatory approach, 224–26
regulatory climate, 211
relative rate increases (charts),
209, 210
utility prosperity, 206–7, 207n
WEPCO rate case and early DSM
programs, 23–24
See also Madison Gas and Electric
Company; Wisconsin Elec-

tric Power Company; Wis-
consin Power and Light
Company
Wisconsin Electric Power Company
(WEPCO)
bonus returns, 85, 209n
DSM activity ratios (table), 236
employee incentive, 213, 248
first DSM incentive, 85, 211–12,
248–49
incentive claim and subsequent
audit, 211–12
rate case (1986), 23–24, 211, 248
ratebased DSM program, 83, 86
recent DSM developments, 36
revised residential end-use forecast, 212
summary information on DSM
incentives (table), 231
Wisconsin Power and Light Company (WP&L)
DSM activity ratios (table), 237
shared-savings incentive, 214–15
summary information on non-
incentive DSM mechanisms
(table), 233
Wisconsin PSC. *See* Wisconsin Public Service Commission
(Wisconsin PSC)
Wisconsin Public Service Commission (Wisconsin PSC)
and annual rate reviews, 204
capitalization of DSM costs policy, 205–6
common goal with utilities, 210
controversial decisions, 210–11
on DSM and risk, 221
emphasis on utility financial integrity, 206–9
escrow accounting, 205–6
FTY ratemaking, 204–5
involvement in the decision-
making process, 211
regulatory policies, 206–11
return on equity policies, 206,
208–9, 209n
return on investment policy, 206

Wisconsin Public Service Corporation (WPSC), shared-savings
incentive, 214–15
WMECO. *See* Western Massachusetts Electric Company
(WMECO)
Wolcott, David R., 271
on DSM incentive mechanisms,
255–69
WP&L (Wisconsin Power and Light

Company), shared-savings
incentive, 214–15
WPSC (Wisconsin Public Service
Corporation), shared-savings
incentive, 214–15

yearly variations analyses, 66

Ziering, Mark, proposal to eliminate
ERAM, 47–48